FOREWORD

The concept of removing long-lived radioactive wastes from the human environment by disposal in deep geological repositories was developed several decades ago. In the intervening years, research efforts world-wide have increased our knowledge and understanding of how underground disposal systems will function over very long periods of time. Significant progress has been made towards implementation of such facilities. There have, however, also been delays in the disposal programmes of several countries. In recent years, the concept itself has moved closer to implementation, but support is increasingly being voiced for postponement and for more review of alternative waste management solutions. On the other hand, reflections in international groups of experts have repeatedly confirmed the conviction of the waste management community that disposal is ethical, environmentally sound and safe, and that other management options are, at best, complementary to disposal rather than true alternatives. Summaries of these reflections are found, for instance, in the Collective Opinions published by the OECD/NEA in 1985, 1991 and 1995.

At its meeting in April 1998, the RWMC found that, in view of the proven value of the Collective Opinions and of the availability of results from several years of further development and application in site characterisation and performance assessment projects, it is now timely to examine whether the findings of the earlier Collective Opinions, and especially those of 1991, are still valid or whether the findings need to be revised or extended to take account of new developments and experience. Even if the earlier findings are still valid, a new document to address specific technical issues might be useful.

The scope of the present report is thus a review of developments in the field of deep geologic disposal and the management of long-lived wastes allocated to deep geologic disposal, particularly those developments that have arisen since the formulation of the 1991 Collective Opinion.

The primary sources of input to the study are the answers to a questionnaire provided by the participating waste management organisations of the Member countries of the NEA Radioactive Waste Management Committee (RWMC), as well as the EC and the IAEA. Additional input came from discussions within the RWMC and from a literature review covering, in particular, results of specialised thematic workshops, programmes of major conferences, annual NEA Nuclear Waste Bulletins, and reports from international organisations over the last decade.

In order to address the different needs of a wide audience, the review is documented in two reports. The present report is targeted at a more specialist audience, and thus contains a more complete documentation of the review findings. A companion, shorter report summarising the review findings and entitled *Progress Towards Geologic Disposal of Radioactive Waste: Where Do We Stand?* is targeted at the wider community of decision makers.

This report was prepared on behalf of the RWMC by a group consisting of C. Combie, C. Pescatore, P. Smith and A. van Luik. Additional assistance was provided by B. Rüegger.

3

CONTENTS

EXECUTIVE SUMMARY

Most nations that generate nuclear power are moving towards "completing the nuclear fuel cycle" through radioactive waste management programmes that ultimately aim to emplace long-lived wastes in a geologic disposal facility, i.e. in a repository deep underground in a suitably chosen rock formation. This approach was selected after considerable debate and discussion. For example, approaches such as sub-seabed disposal, deep-well disposal, and even space disposal, have been discussed and have been found wanting in terms of cost or risk, or impracticable because of political or legal restrictions. The current debates within waste management programmes do not, in many cases, question geologic disposal as the preferred means of disposal for long-lived wastes. Rather, the questions raised are when and where disposal should take place, bearing in mind the need to fulfil ethical obligations, to reduce present and future risks, to ensure that other management options are given due consideration and to enhance wide societal participation in, and ultimately, acceptance of, the disposal strategy.

No nation has ever rescinded its decision to pursue geologic disposal, but some have delayed their repository siting programmes or questioned the wisdom of a selected location. Nevertheless, opinions expressed by a wide cross-section of waste management experts within the scope of the present review confirm the consensus view that geologic disposal is the only feasible route for assuring permanent isolation of long-lived wastes from the human environment. Confidence in the feasibility of secure and safe deep geologic disposal has been built up by:

- the development of detailed repository concepts in many countries;
- the improved understanding of safety-relevant processes though site characterisation and R&D;
- the demonstration of the safety of repository concepts through the application of rigorous safety assessment methods;
- the independent review of these assessments by national and international groups of experts;
- the development, and in some cases demonstration, of technologies necessary for implementation of deep geologic repositories.

In the last decade, significant further progress has been made in the technical aspects of geologic disposal. In general, the necessary technology for geologic disposal is available and can be deployed when public and political conditions are favourable. It is recognised, however, that there is relatively little experience in the application of some of these technologies and therefore demonstration and testing will continue and further refinements will be made. Technical progress has been facilitated through the better integration of the main technical challenges of deep geologic disposal projects, namely the design of engineered systems, the characterisation of potential disposal sites and the evaluation of total-system performance. The interdependencies between these activities are now widely recognised and integrated project-management structures have been established to encourage interaction between different areas.

Progress is evident in the understanding of the natural system, and in the characterisation of potential sites. In particular, advances have been made in measurement methodology and procedures, and in the better appreciation of the heterogeneity (spatial variability) of the system. The improved understanding has made it more feasible to specify technical, safety-related criteria for the acceptability of potential host rocks. In some programmes, partly in response to the greater awareness of the difficulties in fully characterising heterogeneous natural systems, the engineered barrier system has received increased attention in the last decade. In these programmes, there has been a trend towards the adoption of robust engineered systems in order to offset some of the difficulties inherent in assessing the performance of natural systems and thus increase overall system confidence. Robust engineered components should be simple and conservative in their design and must be chemically compatible with, and functionally complementary to, the natural systems. The current emphasis on enhanced assurance through engineering should not, however, detract from the fact that the geologic environment remains a key component of the overall deep disposal system. In fact, for programmes that are considering relatively homogeneous geological formations, the role of the engineered structures may be mainly to provide safety in the operational phase of the repository life.

Progress is also noted in the performance assessment area, particularly in the methodologies for repository systems analysis. Important advances have been achieved in the understanding of the performance of system components and their respective roles, in the quantitative modelling of their behaviour, treatment of uncertainty, in the presentation of assessment findings and in feedback to site selection, characterisation and repository design. As more data from site-characterisation have become available, formal methods to integrate these large and varied datasets into performance assessment models have been developed. Much effort has been focused on increasing the reliability of the methods developed and it is encouraging to observe the increased understanding of the requirements for, and the approaches to, what is referred to, in the technical literature, as confidence building/validation. In particular, it is appreciated that repository development is a step-wise process punctuated by decisions. The evolution of the repository and its environment need not, and cannot, be described completely in performance assessment, given the inevitable uncertainties at any given step. Rather, the technical arguments that support decision making, including performance assessment and arguments that give confidence in its findings, need to be adequate to support the decision at hand. Furthermore, an efficient strategy must exist to deal at future steps with uncertainties that may compromise feasibility and long-term safety.

If technical arguments need to be enhanced in order to support a positive decision, the measures that are taken can include modification to the design and selected location of the repository, to improve the robustness of the concept. They also include measures to increase confidence in the findings of performance assessment, such as improvements to, and testing of, the methods, models and data and the application of more rigorous quality-assurance (QA) procedures for research and development (R&D), assessment decisions and control of input/output datasets. The positive contribution of underground rock laboratories (URLs) is recognised here. Natural or historical analogues are also judged to be useful for providing qualitative understanding of some key processes. It appears, however, that analogues are less amenable to providing the hard evidence for which many had hoped.

Some areas of performance assessment are still being actively developed. These include the treatment of inadvertent human intrusion and the evolution of the surface environment or biosphere, both of which are often handled in a stylised or simplified manner. Confidence in the findings of a performance assessment need not be compromised by the use of stylised or simplified treatments provided that the documentation clearly acknowledges that simplifications have been made and that, due to the presence of irreducible uncertainties, the results of the assessment are to be viewed as indicators of system behaviour, rather than as predictions of consequences that will actually occur in the future. Indeed, this approach is being increasingly advocated in national regulations; if results of safety assessments are being calculated for comparison with regulatory criteria, then the regulator and other relevant decision makers must judge whether a stylisation is acceptable or not. The conservation of information regarding the presence of a repository over a prolonged period can contribute to minimising the likelihood of inadvertent disturbance of the repository by human intrusion and other disruptive human actions. Various studies have been undertaken in the past decade on the conservation of information regarding the presence of a repository over a prolonged period.

The disposal of radioactive waste is guided by national regulatory bodies on the basis of national laws and regulations. At the same time, international collaborative efforts have benefited national waste-management programmes. These efforts provide not only a world-wide exchange of technical information and expertise, but also internationally recognised, non-binding safety standards and binding agreements among states. Trends in the areas of national legislation and regulation include: the agreement that very long timescales (greater than 10 000 years) cannot be ignored, but require different treatments at increasing times; the advocacy of the use of safety indicators in addition to calculational estimates of radiation dose or risk; and the embedding of laws or regulations for waste disposal in a wider environmental regulatory framework. There is, however, currently little direct attention, at the legislative level, to the need for broad-based approaches that assure a uniform degree of environmental protection for society. For example, there is often little linkage between regulation of materials that are radioactive (whether or not they arise from the nuclear industry), and of non-nuclear hazardous or toxic waste forms.

Over the past decade, some real progress has been made towards implementation of repositories. In Germany, a deep repository for low- and intermediate-level waste has been implemented (although operation is currently interrupted), in Sweden, Finland and Norway, geologic repositories at intermediate depth for low- and intermediate-level waste are now operating and, in the USA, a deep geologic repository for long-lived waste commenced operation in March 1999. There is an acknowledgement of the technical and societal value of a stepwise approach to the planning, licensing and implementation of repositories. There have, however, also been delays and there is widespread recognition, within the technical community, that the critical path towards implementation of disposal facilities is increasingly determined not by technical issues, but by the need for public confidence in the concept. Public attitudes towards a disposal project in a particular country are influenced by many factors, including the success of comparable disposal projects elsewhere. The success of disposal projects in neighbouring countries may be particularly influential. The following lessons can be drawn:

- There is a need to demonstrate and communicate to a wider public the consensus and confidence that exists within the waste management community in the concept and technical feasibility of deep geologic disposal.

- External peer reviews and internal self assessments will continue to provide the technical community with valuable means for developing confidence in studies of repository feasibility and safety and for developing a "safety culture" within waste management organisations. In addition, willingness to undergo peer review enhances the reputation of an organisation for openness.

- In order to promote and communicate confidence in geologic disposal to a wide audience, it is necessary to openly discuss the pros and cons of longer-term monitoring, reversibility and retrievability, and to be willing to again evaluate the case for geologic disposal vs. the case for other suggested waste management options.

- In particular, the pros and cons of extended surface storage and of partitioning and transmutation should be objectively debated, since these two strategies have a strong body of support today.

- Independently of the future use of nuclear power in the world, there is a clear need for the development of deep repositories. The large quantities of wastes existing today in civilian and military programmes must be disposed of in a safe manner. Nevertheless, the debate on disposal is inextricably linked to the continuing discussion on future strategies for supply of energy from nuclear and other sources.

- There is a need to view waste management in a wider societal context. In particular, issues such as sustainability, equitable distribution of potential risks and economic reality, are likely to become increasingly prominent, in response to heightened international awareness.

- The adoption of discrete, easily overviewed steps that allow feedback from the public and/or their representatives would promote the strengthening of public and political confidence in the safety of a facility and trust in the competence of the regulators, as well as the implementers, of a disposal project.

The procedures and methods adopted by the waste management community to address its future needs will be nation-, or programme-specific. The development of these procedures and methods, and progress in repository development, will, however, be influenced by developments elsewhere, and will proceed most effectively through the exchange of ideas internationally. It is therefore to be expected that international fora will continue to be important in meeting the future needs of waste management organisations. The sharing of insights and resources in co-operative projects has proved valuable to both implementers and regulators. Dialogue and co-operative projects involving both types of organisation, at the international level, have the potential:

- to demonstrate the wide consensus that exists at the technical level;

- to optimise use of technical and financial resources;

- to clarify understanding of key concepts in repository development;

- to ensure that the process of repository development is fair - and is perceived as being fair also by those outside the waste management community;

- to work towards common understanding of regulatory requirements across different types of waste materials and environmental risks;

- to (at least) rationalise differences between national regulatory guidelines.

International fora allowing cross-party dialogue and co-operative projects are thus likely to continue to play an important role in the future for all those involved in waste management. Even the most encompassing form of international co-operation, sharing a common international repository in a volunteer host country, is a discussion topic which has become more common in recent years.

Broad conclusions reached at the end of this review are that:

- Deep geologic disposal concepts have made significant progress in the past ten years, most especially in the technical areas concerning the understanding, characterisation and quantitative modelling of the natural and engineered safety-barrier systems.

- No radical changes in strategy or in applied methodologies have proven to be necessary. Although, refinements are still being made, deep geologic disposal is effectively a technology that is mature enough for deployment.

- In many programmes, more emphasis is being placed upon the contribution of the engineered barriers, but the natural or geologic barriers in a deep repository continue to play a crucial role in determining the achievable long-term safety.

- All national programmes continue to support deep geologic disposal as a necessary and a feasible technology, even though some countries wish to postpone implementation of repositories or to evaluate other options in parallel.

- There is a general common trend towards advocacy of prudent, stepwise approaches at the implementational and regulatory level to allow smaller incremental steps in the societal decision making process. Discrete, easily overviewed steps facilitate the traceability of decisions, allow feedback from the public and/or their representatives, promote the strengthening of public and political confidence in the safety of a facility along with trust in the competence of the regulators and implementers of disposal projects.

- Although one deep geologic repository, purpose-built for long-lived waste, is now operating, the timescales envisioned ten years ago for the development of deep geologic repositories were too optimistic. The delays that have occurred are partly due to operational causes, but mainly reflect institutional reasons, in large part associated with insufficient public confidence.

- There is an acute awareness in the waste management community of this lack of public confidence; efforts are needed by both implementers and regulators to communicate effectively to decision makers and the public their consensus view that safe disposal can be achieved.

- The implementers and regulators are more willing than ever to heed the wishes of the public in so far as these do not compromise the safety of disposal facilities. One common goal is to establish strategies and associated procedures that allow long-term monitoring, with the possibility of reversibility and retrievability. A number of programmes now consider these issues explicitly.

Alternative means of radioactive waste disposal have often appeared to have promise prior to a thorough consideration of all aspects of the proposal. Several exotic options were studied earlier, and are no longer seriously considered. There are those who, for a variety of reasons, strongly advocate extended surface storage or partitioning and transmutation. The waste management community does not, however, regard extended or "indefinite" surface storage as a real alternative to geologic disposal; at best it offers a postponement of final disposal. Partitioning and transmutation is also not regarded as an alternative; at best it reduces the volume, or changes the isotope distribution, of wastes requiring deep disposal.

11

In summarising the status of the concept of deep geological disposal, it can be clearly stated that real progress has been made, and is being made, thanks to the extensive efforts of numerous experts working in diverse disciplines within national and international waste management programmes. Technical advances and improved societal interactions have taken longer than had been hoped and lengthy delays have occurred in the implementation of deep repositories. One disposal facility has, however, begun operation and a few deep disposal facilities are nearing the point when they will commence operation, although most such facilities are still many years away from implementation. The path towards their eventual implementation may well be eased by the marked increase in public confidence that is to be expected when the first deep repositories are in successful operation.

1.   INTRODUCTION

The study – key questions, aim and scope

In order to evaluate the current status of high-level radioactive waste disposal programmes in its Member countries, to monitor the evolving technical consensus and to guide future activities, the Radioactive Waste Management Committee (RWMC) of the NEA, at its annual meeting in March 1998, agreed to support a *critical review of the evolution of the case for geologic disposal of radioactive waste in the last decade*.

The *key questions* that the review was intended to address were:

- Is progress being made towards offering society a technically safe and suitable means for the management of long-lived radioactive waste?

- In particular, how has the technical case for geologic disposal progressed in the last decade and how has the acceptability of geologic disposal changed?

- What are the key current issues and concerns that will determine future developments and acceptability?

Thus, this report reviews the status and trends in deep geologic disposal in the *areas* of:

- technical developments;

- legislation and regulation;

- implementation and its relationship to public confidence and political acceptance;

with the *aim* of summarising the experiences and current views of participating international bodies and NEA Member countries on technical progress and on the current status of their geologic disposal programmes by:

- highlighting progress made, and setbacks experienced, during the last decade;

- identifying significant issues that have been solved, or raised but not yet solved;

- identifying key issues occupying the radioactive waste management community today.

The *scope* of the review is to provide a realistic summary and evaluation of developments among participating nations in the field of deep geologic disposal, and the management of long-lived wastes allocated to deep geologic disposal. The review concentrates on developments since 1990, i.e.

since the formulation of the 1991 Collective Opinion, referred to in Chapter 2. It focuses on *internal* developments, directly in the hands of the waste management community, i.e. technical, implementational, and regulatory aspects. However, some mention is also made of *external* developments, e.g. financial, political, and public-perception aspects, which influence the progress of disposal projects.

The review draws on input provided by programme managers, regulatory specialists and policy makers, representing a number of international bodies and NEA member countries, and on a survey of the relevant literature. Its findings are intended to help the waste management community in general, and particularly the NEA, to look ahead to the next decade and to propose constructive actions and initiatives. The review is aimed at a wide audience of technical managers and decision makers. For this reason, and also because of the clear recognition in the RWMC of the growing importance of wider issues affecting public confidence in disposal, the review goes beyond purely technical issues.

Input and structure for the review

The primary source of input to the study has been the answers to a questionnaire provided by waste management organisations represented in the NEA/RWMC including the EC and the IAEA. The questionnaire and the list of the organisations that responded to the questionnaire, are given in Appendix 1. The responses to the questionnaire give a detailed insight into developments in many relevant areas. Additional input has come from a review of relevant literature covering, in particular, results of specialised thematic workshops, programmes of major conferences, annual NEA Nuclear Waste Bulletins and reports from international organisations over the last decade, as well as discussions of earlier drafts within the RWMC.

The information provided by questionnaire responses and other sources has been analysed and structured to provide a summary of the key information:

- Chapter 2 summarises the confidence in the concept of geologic disposal that has been acquired by the waste management community.

- Chapter 3 describes progress in:
    - the various technical areas relevant to geologic disposal;
    - the areas of legislation and regulation;

    and describes both progress and delays in:
    - the implementation of geologic repositories.

- Chapter 4 lists the lessons to be learnt, particularly regarding the need to achieve wider public confidence in the geologic disposal concept. It then goes on to examine the role of international fora and co-operation in addressing future needs effectively, and to consider the role of the NEA/RWMC in this context. The Chapter finishes with a concise overview of the principal conclusions that can be drawn from the current review of the status of geologic disposal.

The overview text of the main report is complemented by Appendices giving more detail of the responses obtained from the many respondents to the comprehensive RWMC questionnaire.

2.   CONFIDENCE IN THE CONCEPT OF GEOLOGIC DISPOSAL

Most nations that generate nuclear power are moving towards "completing the nuclear fuel cycle" through radioactive waste management programmes that ultimately aim to emplace long-lived wastes in a geologic disposal facility, i.e. in a repository deep underground in a suitably chosen rock formation. This approach was selected after considerable debate and discussion. For example, approaches such as sub-seabed disposal, deep-well disposal, and even space disposal, have been discussed and have been found wanting in terms of cost or risk, or impracticable because of political or legal restrictions. The current debates within waste management programmes do not, in many cases, question geologic disposal as the preferred means of disposal for long-lived wastes. Rather, the questions raised are when and where disposal should take place, bearing in mind the need to fulfil ethical obligations, to reduce present and future risks, to ensure that other management options are given due consideration and to enhance wide societal participation in, and ultimately, acceptance of, the disposal strategy.

The concept of permanently removing radioactive wastes from the human environment by disposal in deep geologic repositories was developed several decades ago.[1] After considerable internal debate and discussion, a wide consensus has developed within the waste management community that adequate safety and security can be provided for all future times by properly sited and designed repositories and that geologic disposal represents the only feasible route for assuring the permanent isolation of long-lived wastes from the human environment. Geologic disposal is also accepted by this community as an ethical undertaking that should be pursued now, and not left to future generations. This is reflected in the past Collective Opinions of the RWMC (Appendix 7), the content of which can be briefly paraphrased as follows:

1985 Geologic disposal can provide long-term safety.
1991 Methodologies exist for adequately assessing the long-term safety of geologic repositories.
1995 Geologic disposal is ethically and environmentally justifiable.

Although the earlier formulations may appear, with hindsight, rather optimistic, the basic conclusions were re-affirmed in the findings of the first IPAG report (NEA, 1997), as approved by the RWMC, which stated that "... the findings of the NEA/IAEA/CEC Collective Opinion document (of 1991) remain valid".

National debates on the approach to disposal, supported, in some cases, either by a generic environmental impact statement that evaluates alternatives, or by studies that evaluate the feasibility,

---

1.   Practical and ethical considerations were debated early on: e.g. Scott K. G. (1950), "Radioactive waste disposal – how it will affect man's economy?" In *Nucleonics, Vol. 6, No. 1*; Hatch L. P. (1953), "Ultimate disposal of radioactive wastes", In *American Scientist, Vol. 41, No. 3*; and the landmark report published by the National Academy of Sciences of the United States in 1957: "The Disposal of Radioactive Waste on Land".

security and safety of specific geologic disposal concepts, are on-going in many countries. Occasionally, attempts are made to resolve the debate by legislation. An illustrative case is the USA, where a generic environmental impact statement[2] discussed a number of very diverse alternatives, but, in the end, recommended geologic disposal. This study, funded by the United States government, then supported the national decision to implement geologic disposal, which was documented in the Nuclear Waste Policy Act of 1982 and its amendments.

No nation has ever rescinded its decision to pursue geologic disposal, although some have delayed their decision or questioned the wisdom of a selected location. Support for geologic disposal has sometimes been explicitly announced, e.g. by the UK in its White Paper of 1995[3] "affirming in favour of deep disposal rather than indefinite storage", by Canada in its 1996 "Policy Framework for Radioactive Waste", and by Spain in its National Radioactive Waste Management Plans (latest in 1994). Sometimes support is demonstrated less directly, for example, by investment in comprehensive disposal programmes, often including large underground rock laboratories. On occasion, the support is implicit in recommendations, e.g. the 1998 recommendation of the EC to its members to "continue activities on siting, construction, operation and closure of HLW (high-level waste) repositories". Often the continuing confidence in the concept of geologic disposal is expressed by a willingness to proceed with the national efforts of policy makers, regulators and implementers. A final indicator of continued support for geologic disposal is that repeated statements have emerged over the past years from regulators, implementers and oversight bodies to the effect that deep disposal can provide the needed level of safety.

A review of responses to the NEA questionnaire clearly indicates continued support for geologic disposal among all respondents, although, in the case of France, legislation requires that disposal, storage and transmutation are studied in parallel. The Netherlands has been cited by opponents of disposal as an example of a country that has rejected geologic disposal; but this is put into perspective in the responses. The Netherlands also concurs with the general opinion that the preferred long-term option for waste management is disposal in deep geologic repositories and that these can provide the necessary levels of safety, but it requires that underground repositories be designed such that waste can be retrieved if this is deemed necessary at some future time.[4] Even in countries where strong anti-nuclear forces have created legislation, or powerful initiatives for the phase-out of nuclear power (e.g. Sweden, Germany, Switzerland), it is widely accepted that geologic disposal will be necessary.

---

2. *Final Environmental Impact Statement – Management of Commercially Generated Radioactive Waste*, DOE/EIS-0046-F, United States Department of Energy, Washington, D.C., 1980.

3. Review of Radioactive Waste management Policy. Final Conclusions. Cm 2919, July 1995.

4. The issue is, thus, how easy retrieval should be at a given time, given that, even in the extreme case of retrieval from a sealed repository, engineering procedures may be costly but not impossible, and somewhat analogous to the extraction of toxic mineral ores. See also chapter 4.

3.   PROGRESS IN GEOLOGIC DISPOSAL

From the early nineteen-seventies onwards, programmes to develop the concept of geologic waste disposal were initiated in a number of countries and today most NEA Member countries have geologic disposal programmes for long-lived radioactive waste (see, for example, the list of respondents to the RWMC questionnaire given in Appendix 1). Considerable resources have been invested in making the path to radioactive waste disposal a reality.

In the last decade, significant further progress has been made in the technical aspects of geologic disposal (disposal technology, understanding of the natural system, site characterisation, understanding of the roles of the natural and engineered barriers, the assessment of, and demonstration of confidence in, long-term safety and understanding of the role of information conservation in deterring human intrusion) and in the areas of legislation and regulation.

In this period, however, there has been little *fundamental* change in policy, in implementing or regulatory structures (Appendix 2), or in technology. The lack of fundamental change does not imply that the progress that has been made is unimportant; rather, it indicates that the key decisions taken more than ten years ago have not needed radical revision.

Technical progress

Technical progress has been facilitated through the better integration of the main technical components of deep geologic disposal projects, namely:

- design of disposal systems;

- characterisation of potential disposal sites; and

- evaluation of total-system performance.

The interdependencies between these activities are now widely recognised, and integrated project management structures have been established to encourage interaction between different areas. In particular, a more integrated strategy of site characterisation and design is now widely followed, in which experimental and design work is focussed on areas that promise to improve system safety and confidence.

The following points summarise the principal technical developments.[5]

---

5. An overview of questionnaire responses on the status and trends with respect to technical developments in deep geological disposal is given in Appendices 2, 3, and 4 augmented with input from other referenced documents. It is noted that, although there were nuances in judging the completeness of the technology and safety methodology, there was broad consensus on all key issues.

Disposal technology

In general, the necessary technology for geologic disposal (for waste conditioning and repository design and engineering) is available and can be deployed when public and political conditions are favourable, although there is relatively little experience in the application of some of these technologies and therefore demonstration and testing will continue and further refinements will be made.

Progress in the areas of waste conditioning and repository design and engineering has been achieved mainly by gradual improvements in design concepts and in the technology that is available to implement them. There have been relatively few new design concepts and technological breakthroughs; the trend has been towards incremental improvement and to more rigorous demonstration of existing concepts. There is increasing emphasis on demonstrating the feasibility of reliable fabrication and also on assuring long-term performance. The conclusion that the necessary technology for geologic disposal is now available and can be deployed when public and political conditions are favourable has been drawn by numerous proponents. It is also indirectly supported by the widespread recognition that problems of a more sociological nature are today on the critical path towards repository implementation.

Understanding of the natural system and site characterisation

Definite progress is evident in the understanding of the natural system, and in the characterisation of potential sites, particularly concerning measurement methodology and procedures and in the better appreciation of uncertainties and the heterogeneity of the system.

Techniques for the measurement of data have continued to be refined and tested at potential sites and at underground research laboratories, and the performance of processing and interpretation methods and codes has improved, such that field data are able to yield more relevant information than was previously the case. Experience has, however, shown that non-conclusive data can emerge, making it necessary to collect still more data in order to understand the system – i.e. to reduce the uncertainty in a system that is more complex than originally believed. Heterogeneity has become increasingly recognised as a universally present feature of the geologic environment. Recent emphasis in site characterisation has been on the needs of safety assessment and, in particular, on the understanding of water-conducting features and the flow of groundwater through them. This is generally considered to provide the most likely means by which any radionuclides released from a repository might be conveyed to the surface environment. Other aspects of characterisation that are important to safety assessment, such as hydrogeochemical analysis and the analysis of the roles of colloids, organics and potential microbial processes, have also received attention.

Advances have been made in the specification of safety-relevant technical criteria for the acceptability of potential host rocks (including, in the case of Sweden, acceptance criteria for individual deposition holes), although less attention has, up to now, been paid to aspects of site characterisation that are required in order to optimise repository design.

Technical criteria for the acceptability of a potential host rock are valuable, particularly if developed in advance of an exploration programme, in that they not only serve to guide that programme, but may favour public acceptance, by demonstrating transparently that an implementer would be prepared to abandon a site, should it prove unacceptable. On the other hand, if the criteria are unrealistic or over-simplistic, they may lead to the abandonment of a site that is, in fact, adequate in terms of its ability to assure public health and safety.

In summary, improved understanding of the natural system has enabled site characterisation to develop from a relatively unstructured collection of geologic data into a focused technical activity, aimed at gathering the key data required to assess the performance of a disposal system.

Role of the natural and engineered barriers

In some programmes, partly in response to the greater awareness of the difficulties in fully characterising heterogeneous natural systems, the engineered barrier system has received increased attention in the last decade. In these programmes, there has been a trend towards the adoption of robust engineered systems in order to offset some of the difficulties inherent in assessing the performance of natural systems and thus increase overall system confidence. Robust engineered components should be simple and conservative in their design and must be chemically compatible with, and functionally complementary to, the natural systems. The current emphasis on enhanced assurance through engineering should not, however, detract from the fact that the geologic environment remains a key component of the overall deep disposal system. In fact, for programmes that are considering relatively homogeneous geological formations, the role of the engineered structures may be mainly to provide safety in the operational phase of the repository life.

Prior to site characterisation, modelling of site performance may have to be based on a rather idealised view of natural system properties, and may not reflect the realities and uncertainties understood after characterisation. Typically, this means that when more is known about a site, more is also known about heterogeneity and uncertainty. In the last ten years there have been significant advances in the analysts' ability to understand and characterise heterogeneous systems. It has also been recognised that a robust engineered system, that is compatible with, and functionally complementary to, the natural system, can help to counteract residual uncertainty in the natural system, particularly in the case of a repository sited in a fractured hard rock. The natural system protects the engineered system from human intrusion and from the rapid changes in environmental conditions experienced at and near the ground surface. It also provides an environment favourable to the longevity of the materials of which the engineered system is made. It provides a chemical and physical buffer that is conducive to slow container degradation and very slow radioactivity releases thereafter. Thus, the vast majority of radionuclides in the emplaced inventory, in a practical sense, never leave the location at which they are emplaced. It is only relatively few radionuclides that are potentially released from a well-designed engineered system, and these are diluted through mixing with groundwater as they travel away from the repository. Many are also retarded during their transport through tortuous pathways and undergo significant decay during transport. Nevertheless, the current emphasis, in some programmes, on enhancing assurance through engineering should not be allowed to detract from the fact that the geologic and hydrogeologic setting of the repository remains a vital part of the overall deep disposal system, and the uncertainties associated with engineered systems should not be overlooked.

Assessment of, and confidence in, long-term safety

Significant progress is noted in the performance assessment area, particularly in the methodologies for repository systems analysis and in the development of more defensible models that reflect progress made in both natural and engineered system understanding and design.

Aspects of performance assessment where particular progress has been made include the understanding of the performance of system components and their respective roles, the quantitative modelling of their behaviour, treatment of uncertainty, the presentation of assessment findings and feedback to site selection, characterisation and repository design (each area is discussed in detail in Appendix 3). The more sophisticated use of probabilistic codes has also been noted and there has been progress in understanding the strengths and weaknesses of probabilistic assessment techniques with respect to a deterministic approach (NEA, 1997). Some programmes now implement a combination of both methods.

In addition, in response to the need to deal with large site datasets, progress has been made in more formal methods of data reduction for use in assessment models. Although the increased use of data from actual sites (and the increased detail in the specification of repository design) has presented new challenges and requires more resources than expended in earlier performance assessments, no insurmountable problems in the use of performance assessment methods have been encountered.

Concerning confidence building/validation in performance assessment, many perceive that there has been an increase in understanding. It is widely acknowledged that it is impossible to describe completely the evolution of an open system, such as a repository and its environment, that cannot be completely characterised and may be influenced by natural and human induced factors outside the system boundaries. As discussed below, however, a complete description is not a requirement for decision making in repository development.

The role of performance assessment in the decision making process for repository development

In general, repository development proceeds by an incremental and flexible stepwise approach. The decision whether or not to advance from one step to the next, if it is the responsibility of technical specialists and managers in the implementing and regulatory bodies, requires technical arguments that give confidence in the feasibility and long-term safety of the proposed concept. The depth of understanding and technical information available to support decisions vary from step to step.

It is appreciated that decision making requires only that the technical arguments, including performance assessment and arguments that give confidence in its findings, are adequate to support the decision at hand, and that an efficient strategy exists to deal at future stages with uncertainties that may compromise feasibility and long-term safety.[6]

Methods to enhance confidence in the safety indicated by performance assessment

If technical arguments need to be enhanced in order to support a positive decision, the measures that are taken can include modification to the design and selected location of the repository, to improve the robustness of the concept. They also include measures to increase confidence in the findings of performance assessment, such as improvements to, and testing of, the methods, models and data and the application of more rigorous quality assurance procedures for research and development, assessment decisions and control of input/output datasets.

The positive contribution of underground rock laboratories (URLs) is recognised here; natural analogues are also recognised as useful, although less amenable to providing the hard evidence for which many had hoped. There has been increased emphasis on the testing of key performance assessment models and databases in the work programmes at URLs, as part of the confidence-building process (see Appendix 3). Natural analogues provide a further component of the confidence building process, although enhanced integration of natural analogues into performance assessments is still seen as an area warranting further development.

---

6. *Confidence in the Long-term Safety of Deep Geological Repositories: Its Development and Communication.* OECD/NEA, Paris 1999.

Some uncertainties can be identified that must be addressed in performance assessment, but are, in practice, impossible to quantify and to reduce (e.g. inadvertent human intrusion and the evolution of the surface environment or biosphere). Such uncertainties are often treated in a stylised or simplified manner.

Confidence in the findings of a performance assessment need not be compromised by the use of stylised or simplified treatments provided that the documentation clearly acknowledges that simplifications have been made and that, due to the presence of irreducible uncertainties, the results of the assessment are to be viewed as indicators of system behaviour, rather than as predictions of consequences that will actually occur in the future. If results for comparison with regulatory criteria are being calculated, then the regulator and other relevant decision makers will judge whether a stylisation is acceptable or not.

The role of information conservation in deterring human intrusion

Theoretical studies have been undertaken on the conservation of information regarding the presence of a repository over a prolonged period. Practical steps have also been taken that may lead to the establishment of radioactive waste archives.

The conservation of information regarding the presence of a repository over a prolonged period can contribute to minimising the likelihood of inadvertent disturbance of the repository by human intrusion and other disruptive human actions. These issues and, in particular, the value of markers and archives, have been discussed by an OECD/NEA Working Group, with the findings published in 1995.[7] The issues have also been considered in a Nordic Study in 1990, in the licensing of the WIPP facility in the USA and by the IAEA.[8] In some countries, regulations are already in place that require records describing the wastes designated for disposal in a repository to be conserved in some form of post-closure archive. Furthermore, in the "Joint Convention on the Safety of Spent Fuel Management and on the Safety of Radioactive Waste Management" it is foreseen that Member States periodically report their practices regarding new disposal sites. This may represent a first step towards an international archive for radioactive waste disposal.

Progress in legislation and regulation

The disposal of radioactive waste is guided by national regulatory bodies on the basis of national laws and regulations.[9] At the same time, international collaborative efforts have benefited national waste management programmes, providing not only the world-wide exchange of technical information and expertise, but also internationally recognised, non-binding safety standards and binding agreements among states.

Both national and international developments are summarised below.

---

7.   *Future Human Actions at Radioactive Waste Disposal Sites*, OECD/NEA, Paris, 1995.

8.   *Maintenance of Records for Radioactive Disposal*, IAEA, Vienna, 1998 (in press).

9.   The overview of questionnaire responses on the status and trends with respect to legislation and regulation is given in Appendix 2.

Regulator and implementer consensus on key issues

In early 1997, an extremely valuable NEA Workshop took place in Cordoba,[10] involving active regulators and implementers from a wide range of national programmes. The first novel feature of the meeting was that it was explicitly dedicated to encouraging dialogue between these groups. One aim was to establish where real consensus on key issues existed, in order to avoid confusing a wider audience with debate on apparent differences which might be based more on semantics than substance. The participants examined waste disposal objectives and criteria, trends in performance assessment and the conduct of regulatory processes. In the specific area of open regulatory issues, the key points agreed upon were that:

- A more sophisticated yardstick than only the "high-level" criteria of dose or risk is needed.

- The waste management guiding principle that adequately protecting mankind also protects the broad environment was questionable.

- Long timescales (greater than 10 000 years) cannot be ignored, but require different treatments at increasing times.

- Communication between regulators and implementers, as well as with broader groups, needs improvement.

- Licensing requirements ("rules of the game") should be framed well ahead of when they are to be applied – and can justifiably include simplified or stylised approaches.

- The stepwise approach is valuable for implementers, regulators, decision makers and the public.

All of these conclusions are also clearly reflected in the responses offered by representatives of NEA Member countries to the present questionnaire.

Developments in internationally recognised safety standards

The International Commission for Radiological Protection (ICRP) formulates fundamental radiation principles and criteria for world-wide application. Since the beginning of the decade, the ICRP has established radiation protection principles for a number of practical situations where exposure is not foreseen but at the same time cannot be excluded, and has elaborated its general guidance[11] on policy and ethical considerations relevant to radioactive waste management.[12] The ICRP has also issued advice that is specifically relevant to the performance assessment of repositories for long-lived waste, e.g. the problems associated with a changing biosphere, the rationale for the exclusion of particular scenarios in a safety case presented as part of a license application and the interpretation of the principle that future generations should enjoy the same protection as the present.[13]

10. *Regulating the Long-term Safety of Radioactive Waste Disposal*, Córdoba, Spain, 20-23 January 1997. Proceedings edited and distributed by the Consejo de Seguridad Nuclear, Madrid, 1997.

11. *ICRP Publication 64: Protection from Potential Exposure: A Conceptual Framework. A Report of a Task Group of Committee 4 of the International Commission on Radiological Protection.* Annals of the ICRP Vol. 23/1, 1993.

12. *ICRP Publication 46: Radiation Protection Principles for the Disposal of Solid Radioactive Waste.* Annals of the ICRP Vol 15/4, 1986.

13. *ICRP Publication 77: Radiological Protection Policy for the Disposal of Radioactive Waste.* Annals of the ICRP Vol. 27 Supplement, 1998.

Based on the work of the ICRP, the International Atomic Energy Agency (IAEA) has issued internationally recognised non-binding standards on radioactive waste safety, the Radioactive Waste Safety Standards (RADWASS). These are intended to establish an ordered structure for safety documents on waste management and to ensure comprehensive coverage of all relevant subject areas. The IAEA has, more recently, established a working group to explore and, where possible, develop consensus on issues relevant to deep geologic disposal and the findings, when completed, together with guidance from the ICRP, are likely to be used in developing new safety standards in several nations.

Agreements among states

Over the past decade, several binding agreements among sovereign States have been implemented with the support of the IAEA. These agreements may be viewed as a component of a global framework for fostering intergovernmental collaborative efforts in the area of nuclear safety. Specifically regarding radioactive waste disposal, the "Joint Convention on the Safety of Spent Fuel Management and on the Safety of Radioactive Waste Management" was adopted in 1997 and, when ratified, will be the first legally binding instrument in this area. The implementation of the Convention will formally be pursued through peer review of national reports at Review Meetings of the Contracting Parties.

Trends in national laws and regulations

There is a trend in national regulations towards the advocacy of the use of safety indicators in addition to calculation estimates of radiation dose or risk and of stylised approaches for assessing long-term impacts.

This trend reflects international discussions, and the work of the ICRP and IAEA, regarding the difficulties of demonstrating compliance with radiological and other safety objectives in the far future. As a result, safety indicators other than radiation dose and risk have been proposed, such as environmental concentrations and biosphere fluxes of radionuclides, as a way of broadening the safety case and its appeal to different audiences. In addition, it has been recognised that, although there is no scientific rationale for cutting off repository safety assessments at an arbitrary point in time, the nature of performance assessments over different timescales after closure of a repository cannot be the same. In particular, uncertainties increase with increasing time and so the results of assessments have to be regarded as indicators of safety, rather than as real predictions of impacts. These considerations have brought with them proposals to develop internationally agreed stylised approaches for assessing long-term impacts, e.g. the concept of reference critical groups and biospheres. These proposals are reflected in the documents issued by the IAEA Working Group on Principles and Criteria for Radioactive Waste Disposal. Furthermore, specific guidance on these issues is expected in the forthcoming ICRP report on "Radiological Protection Principles for the Disposal of Long-lived Solid Radioactive Waste".

There is also a tendency towards the embedding of national laws or regulations for waste disposal in a wider environmental regulatory framework.

In many countries, an environmental impact assessment (EIA) must be carried out for facilities whose construction or operation might result in a significant impact upon the environment.[14]

---

14. This includes, for example, those countries that have implemented EC relevant directives; a DG-XI study on environmental impact assessments and implementation of relevant directives will shortly be completed.

The EIA is carried out by the operator or proponent of the facility and made available for public comment. There is, however, little mention of issues like sustainability or of broad-based approaches at the legislative level to assure a uniform degree of environmental protection amongst radioactive waste sources and also between nuclear and non-nuclear hazardous waste forms. Furthermore, although many countries claim coherence between legislation affecting radioactive waste disposal and other environmental legislation, there is currently little direct evidence that objective risk-based regulations are ensuring equitable treatment – and proof of treatment – of all potential hazards to humans. For example, in some countries, permits to allow injection of toxic materials into boreholes require in principle that the applicant demonstrate that the toxins will "never" endanger groundwater, but analyses projecting potential behaviour into the very far future are not required to be submitted and reviewed in an exhaustive manner equivalent to that for radioactive wastes.[15]

Progress and delays in implementation

Over the past decade, there has been increased appreciation, at a political level, that geologic disposal is necessary. All countries acknowledge their responsibility to propose solutions to their own disposal problems and real progress has been made towards implementation.

Throughout the waste management chain there are numerous examples of progress towards implementation of individual steps, all of which lead towards the ultimate objective of final disposal. Each advance not only brings the completion of the cycle closer, it also helps build public confidence by demonstrating that facilities can be sited, implemented and safely operated by the waste management community. Areas where advances have been made include:

- *Implementation of facilities for treating and storing wastes*

  Particularly notable are the developments in dry storage technologies and in the implementation of centralised storage facilities for spent fuel and HLW.

- *Experience in site characterisation*

  Ten years ago, little real site data were available and data collection strategies and methods were much less developed. Today extensive programmes involving detailed characterisation with geophysics, numerous boreholes and even exploratory shafts and ramps have been carried out in various countries. The range of geological, hydrologic and geochemical data which is today available for analysing sites is far wider than was the case in the eighties.

- *Construction of underground rock laboratories*

  These facilities, in addition to providing invaluable test beds for practical methods and theoretical models, also are extremely effective confidence building instruments for all those who can observe their operation. In Appendix 3, the numerous underground test facilities in operation around the world are listed in more detail.

- *Commissioning of facilities at intermediate depth*

  The obvious examples here are the low and intermediate waste (L/ILW) repositories now operating in Sweden, Finland, and Norway. The fact that final disposal of radioactive wastes

---

15. Code of Federal Regulations, Protection of the Environment 40 Part 148, "Hazardous Waste Injection Restrictions". Office of the Federal Register, National Archives and Records Administration, Washington, D.C. 1997.

in underground caverns is already taking place is direct proof of the feasibility of such projects.

- *Imminent implementation of some deep repositories*

The advent of a deep repository accepting long-lived waste will mark a major milestone in geologic disposal. Opponents of the concept point out that no such facilities have been commissioned to date. The earlier responses of the nuclear community – that there is no pressing need or that necessary cooling times lead to later disposal dates – are no longer valid. There are sufficient cooled wastes available in some countries to justify commencing disposal, although, in most cases, the limited volume of the wastes imply that urgent measures are not necessary. In Germany, a deep repository for L/ILW has been operating at Morsleben, although waste implacement has been stop recently. The deep disposal project for long-lived waste nearest to full implementation is that at the Waste Isolation Pilot Plant (WIPP) in the USA. In May 1998, the EPA issued the license to begin disposal operations for TRU waste at WIPP and on 22 March 1999 a US court ruled in favour of the WIPP starting operation. Furthermore, the State of New Mexico is in the process of issuing a license to allow disposal of chemically hazardous TRU waste ("mixed waste"), and it has accepted that some of the waste stored at Los Alamos is not to be considered as mixed waste. Shipment of this TRU waste to WIPP commenced in March 1999. The Yucca Mountain "Viability Assessment" was published in December 1998 for consideration by Congress. It is expected that a site recommendation will be made in 2001 and, if Yucca Mountain is recommended, that a license application could be made in 2002.

It is interesting to examine the views of those within the waste management community on the most positive developments within the field during the past ten years. For this reason, appropriate questions were posed in the questionnaire circulated by the NEA. Table A6.3 in Appendix 6 shows clearly the importance of moving projects ahead in a visible fashion. Advances cited by participants are positive not only in the national programme involved, but also in other countries, testifying yet again to the strong international connections in radioactive waste management.

Of course, these positive judgements are from waste management specialists and not from the public. Moreover, there is also a negative side to developments – mainly centred upon the difficult problem of gaining the public confidence that is necessary in order to advance waste disposal programmes into the phase where specific sites are identified and facilities constructed.

There is an acknowledgement of the value of a stepwise approach to the planning, licensing and implementation of repositories.

The novelty and complexity of the task of repository development mean that the detailed planning, licensing and implementation of a disposal facility in a single step is not possible. Rather, there is a need for an incremental and flexible procedure that allows gradual growth in confidence as information and experience are acquired. A stepwise approach is valuable not only for these objective, technical and procedural reasons, but also because it requires smaller steps in the societal decision making process. Although, in most programmes, a lengthy, stepwise development process is foreseen through generic and site specific investigations up to repository implementation, the meaning of the term "stepwise approach" differs between countries and a clearer definition of what the term means in practice should be developed. Typically, it involves a number of development stages, punctuated by interdependent decisions regarding, for example, interim surface storage, siting and design, safety assessment, site characterisation, the licensing of construction, operation and closure,[16] sealing and post-closure monitoring that are taken throughout the development of a facility. Discrete, easily overviewed steps facilitate the traceability of decisions, allow feedback from regulators or from the public and promote the strengthening of public and political confidence in the safety of a facility and trust in the competence of the regulators, as well as the implementers, of a disposal project. The increased confidence that comes with a stepwise approach can also, to some extent, derive from observations of correct system behaviour over longer time periods ("demonstration activities"). In highly litigative societies, it should be noted, the price paid for a highly formalised or legislated stepwise procedure can be the extention of project timescales and budgets beyond the levels needed to achieve the technical and societal aims.

It is widely recognised that not only acceptance by the technical community, but also confidence on the part of the general public needs to be achieved; a lack of public confidence is a major factor in recent delays, and may prove the most important obstacle on the path to implementing deep geologic disposal.

In spite of the acknowledgement by the waste management community, and by many policy making bodies, that geologic disposal is the preferred route for ultimately isolating long-lived wastes from the human environment, and in spite of the technical progress that has been made, timescales on which most implementing organisations hope or wish to operate deep repositories have increased compared to the situation 10 years earlier. Delays can result from the need for additional technological and scientific development, but are more commonly associated with difficulties achieving public confidence and political acceptance. One example of this is the observation by the Government Panel reviewing Canada's Nuclear Fuel Waste Management and Disposal Concept that technical safety had been demonstrated, but that lack of public confidence warranted the adoption of additional complementary approaches. A critical question for all stakeholders is how to determine when sufficient societal confidence has been achieved. In most countries, interest in waste management issues is concentrated in special interest groups from both sides of the debate, with society at large showing less interest. Without a definition of broad societal confidence, waste management organisations lack a yardstick against which to measure their success in achieving such confidence and governments lack adequate guidelines for making policy.

Specific factors that have resulted in the stretching of timescales include:

- unanticipated delays[17] due to changes in government policy and to regulatory or public review activities (for example, in the UK, Sweden, Switzerland, Germany, Canada and the USA);

- the general move to a phased approach, including a phase with retrievability (e.g. in Sweden, Switzerland and the USA);

- the re-emergence of pressure to examine "alternatives" like partitioning and transmutation (P&T).

---

16. In the USA, only a single licensing action is taken, but a stepwise approach is, nevertheless, followed. A safety case is made to support construction authorisation and this is updated prior to receiving the waste for emplacement, and updated again prior to the license amendment to seal the repository. Throughout this time, there is ongoing research and periodic assessments of performance.

17. A contrast to the tendency towards stretching or slowing programmes is provided by some countries such as the Czech Republic where high-level waste programmes are in the process of being built, with international co-operation making rapid progress possible.

For example, loss of its public enquiry is claimed to have set the UK programme back by "several decades"; loss of a public referendum in Switzerland has delayed the LLW/ILW geologic disposal plan by many years. The ambitious programmes in Sweden and the USA have experienced delays measured in years, and uncertainties in developments have led to postponement of reference deep disposal dates in Canada by years, or even decades.

In Canada, an independent Environmental Assessment Panel has made recommendations pertaining to responsibilities, funding and the preferred approach for the long-term management and disposal of nuclear fuel waste, most of which have recently been accepted by the Government. In all areas, societal confidence aspects should be thoroughly taken into account. In particular, a separate fund should be established for designing and siting facilities according to the preferred approach for nuclear waste management, including geologic disposal. The interesting conclusions of the Panel were that, although a technical safety case had been made for the generic concept, there was too little public confidence to justify moving immediately to a siting phase.

In the UK, following the refusal of planning permission for the deep rock laboratory at Sellafield, a governmental decision is awaited that will recommend, and outline, the next steps to be taken in radioactive waste disposal.

In France, it has been legislated that no final decision on waste management strategy should be taken before 2006, when 15 years will have been spent exploring in parallel the three options of disposal, storage, and P&T. The government has, however, given the decision confirming the plan to develop two underground research laboratories. A new site in hard rock must be sought and proposed within a three-year period by the same volunteer process as used before. A third site (the Gard site) would be considered for research in sub-surface interim storage.

In Germany, despite the huge technical efforts invested at the Gorleben and Konrad sites, the new national government, while accepting the appropriateness of a deep geologic repository as a means of final disposal, is reconsidering the location of such a repository and the timing of the programme.

The Netherlands has legislatively decided to postpone geologic disposal and to support only work based on fully retrievable wastes at all foreseeable future times.

Such delays present national programmes with practical problems regarding, for example, the loss of technical capability and the assurance of adequate funding if funds are not preserved for their intended use over the period of delay. Furthermore, without a means of final disposal, the waste continues to present national problems that, unless resolved, will pass the burden of responsibility to future generations that have not benefited from the nuclear power generated from what is now the spent fuel and reprocessed waste, in contradiction to the principle of intergenerational equity.

Nevertheless, in spite of the delays, no nation has rescinded its decision to pursue geologic disposal and the consensus for pursuing geologic disposal as the only feasible route for assuring permanent isolation of long-lived wastes from the human environment is unaffected.

4. LESSONS TO BE LEARNT – CONCLUSIONS DRAWN

Actions required of the waste management community

Support for geologic disposal within the waste management community is built, in part, on confidence in the technical feasibility of implementing secure and safe repositories. As discussed in Chapter 3, repository concepts have been developed in many countries, the technologies necessary for implementation have been developed, an understanding of safety-relevant processes has been acquired though site characterisation and R&D, and the safety of the concepts has been demonstrated through the application of rigorous safety assessment methods.[18] These assessments have, in many cases, been independently reviewed.

The technical consensus within the waste management community on the merits of geologic disposal is not wholeheartedly shared by the general public and by others in society who influence decision makers. There is widespread recognition, within the technical community, that the critical path towards implementation of disposal facilities is determined not as much by unsolved technical problems, as by lack of public confidence and political acceptance of the concept.

The following lessons can be learnt:

(i) There is a need to demonstrate and communicate to a wider audience the consensus and confidence that exists within the waste management community in the concept and the technical feasibility of deep geologic disposal.

The technical personnel involved in most activities include the regulators and the implementers, between whom a first consensus on disposal safety must be achieved before public confidence and political acceptance can be expected. In particular the following points should be emphasised and communicated outside the community of regulators and implementers:

- There is a clear consensus on the need for geologic disposal within waste management programmes.

- There is high confidence in the technical feasibility and safety of a properly designed and sited repository.

- Clear procedures have been, or are being, proposed for staged siting studies and repository development.

---

18. Many integrated safety assessments of disposal projects are referred to in Appendix 3. Similar methods have been applied in areas in addition to that of geologic disposal; for example, to the radiological situation at the Atolls of Mururoa and Fangataufa, in a study co-ordinated by the IAEA in 1998.

- The most valuable expression of consensus and confidence will be through the successful operation of one or more repositories.

Public confidence in a repository concept in a particular country would no doubt be enhanced by success of comparable disposal projects in other countries, especially in their own vicinity. Events, positive or negative, in one country certainly influence attitudes in others. A negative example is, of course, Chernobyl. The most obvious recent positive example is the national level regulatory compliance certification of WIPP.

(ii)   External peer reviews and internal self assessments will continue to provide the technical community with valuable means for developing confidence in studies of repository feasibility and safety and for developing a "safety culture" within waste management organisations. In addition, willingness to undergo peer review enhances the wider reputation of an organisation for openness.

Many programmes and studies have, in the past decade, undertaken both external peer reviews by teams of international standing and internal self assessments. External peer reviews have been organised, for example, through the NEA and IAEA, and provide:

- confidence, on the part of the organisation requesting the review, in the quality (or otherwise) and relevance of their work and guidance as to the future direction that the work should take;

- confidence, on the part of decision makers, in the openness of the organisation to external scrutiny.

In addition, those participating in the review derive a valuable insight into the approaches and achievements of an organisation other than their own.

Self assessment, within an organisation, can broaden the appreciation, on the part of staff, of the objectives to be achieved and the means for achieving them, particularly with regard to safety. It can also promote good communications within an organisation.[19]

(iii)   There is a need for more public involvement and improved communication. This can be facilitated by discrete, easily overviewed steps in repository planning, licensing and implementation, that allow feedback from regulators and from the public, and promote the strengthening of public confidence in safety and trust in the implementing and regulatory organisations.

Both implementers and regulators are used to putting their technical case to decision makers and even to the politicians. Both recognise and understand that public involvement is now a key issue. Both are in general realistic, or even pessimistic, in their assessment of how successful they have been, or indeed can be, in this sensitive area. In response to this need of national programmes, effective methods for communicating with the public and for gaining public confidence in the development of appropriate national solutions is a key area needing attention. A wide variety of means is available by which the public can give input to the planning and implementation process. Current practice in many countries is to adopt a stepwise approach to repository development. This reflects the clear public aversion to large irreversible steps. In Switzerland, for example, a prime reason for loss of a referendum at the Wellenberg site was that the public would have preferred to give an initial permit only for the construction of an exploratory tunnel, rather than directly for the final repository.

---

19.   IAEA Bulletin, 40/2/1998.

The issue of communication of the case for geologic disposal to the general public and to societal decision makers is discussed in Appendix 6. It is recognised that the success of efforts to date is questionable. Greater public confidence can be gained through:

- communications at different levels, with complete openness being demonstrated at all times;

- a clear decision-making process, with well defined roles of implementers and regulators and with regulatory frameworks that do not contradict one another and that, ideally, take into account the risk basis of other nuclear and non-nuclear regulations;

- making full use of the diverse procedures by which public input can be integrated into the planning and development of repositories.

One hope of waste management organisations is that the public may ultimately gain confidence through the demonstrated willingness of implementers to take all their appropriate concerns seriously. The public has proved its ability to block (or certainly cause detailed re-evaluation of) proposed nuclear developments in several countries, and, without some degree of national public acceptance, repository projects are not likely to succeed.

(iv)   In order to promote and communicate confidence in geologic disposal to a wide audience, it is necessary to openly discuss the pros and cons of longer-term monitoring, reversibility and retrievability, and to be willing to again evaluate the case for geologic disposal vs. the case for other suggested waste management options.

Although the vast majority of the waste management community believes geologic disposal to be an ethical course which can provide a permanent solution to the waste problem, there has arisen an alternative ethical argument that it is preferable to leave future generations with as many options as possible. Therefore it is argued that disposal – which is seen by much of the public and by some societal decision makers as a strategy which will reduce future burdens at the expense of removing flexibility – should be either postponed, or implemented in a manner ensuring that monitoring and, if necessary, reversibility and retrievability are maintained.

Reversibility and retrievability are, in practice, being increasingly considered by waste management organisations. Box 1 summarises the intent of these procedures and gives examples of recent national developments in the area.

Despite the confidence in geologic disposal on the part of the waste management community, there is, nevertheless, a continuing pressure to identify and assess other possible waste management routes.

The recent review of the Canadian programme by an independent panel, for example, recommended a study of options for spent fuel management other than the geologic disposal concept that was the subject of the review. Several "exotic disposal routes", such as disposal into space, sub-oceanic subduction zones or polar ice, were already fairly extensively studied in the seventies and are no longer the subject of serious consideration, at least within the community. Sub-seabed disposal has also been evaluated (e.g. in the PAGIS study of the EU) and found to be promising from the point of view of safety. Implementation of sub-seabed disposal would, however, depend on international acceptance and development of an appropriate international regulatory framework. Currently, most countries would interpret their obligations under the London Dumping Conventions as precluding the disposal of nuclear waste under the seabed. The most common management alternatives currently raised in public debate on waste disposal, as well as by implementers and regulators striving for completeness in the range of options considered, are extended or indefinite storage and partitioning and transmutation of long-lived radionuclides in wastes. The study of possible waste management options is further reviewed in Box 2.

Box 1. **Monitoring, reversibility and retrievability**

Reversibilty can be an aim throughout waste management. Wastes should not, as far as possible, be conditioned into a form which precludes taking advantage of future technological developments; sites should not be definitively nominated until options have been explored; designs should not be frozen too soon; commissioning, operation and closure of a repository should all be a process of small steps each of which allowing the consequences of the commitment to be fully considered. Reversal of the entire disposal process by retrieving previously emplaced wastes is perhaps the most challenging and contentious issue. At a practical, technical level easy retrievability can conflict with maximal isolation; in the area of disposal economics, planning for possible retrieval leads to the need for financial provisioning for subsequent alternative disposal technologies. Nevertheless, in view of the increasingly widespread desire to keep future options open, there is a corresponding increased interest in evaluating retrievability or reversibility for a (sometimes long) period of time, during which monitoring either continues to confirm safety, or brings previous safety evaluations into question.

For long-lived waste governments have either recommended, or require consideration of, retrievability or reversibility in the Netherlands, France, and the USA (for a limited time after emplacement). Implementers in countries like Sweden, Switzerland, Canada and the UK have read the signs given by the public and politicians, and have themselves built in features assuring longer term retrievability than originally envisioned. The periods explicitly considered run through the whole operational period and over a period in the order of a hundred of years beyond.

The question of reversibility and retrievability, with its ethical, sociological, security, safety and technical aspects, certainly represents, in the words of the EC respondent to the NEA questionnaire, "an issue which has aroused increased attention" in recent years, and it would justify a clear statement of its implications by the international community.

Although it is ready to examine these issues in order to fulfil public or political wishes, the waste management community in general does not believe storage to be an alternative to (but rather a means of postponing) disposal and recognises that extended storage has its own environmental and social consequences, relying, for example, on societal stability to ensure continuing safety. A decision on long-term storage must balance the risks inherent in such a policy, against any possible benefits.

The community is, furthermore, often sceptical about the practicality of options such as partitioning and transmutation (P&T), which still leave a waste stream requiring geologic disposal. In particular, although P&T has the potential to reduce, though not eliminate, the inventories of some longer-lived radionuclides, there exist significant uncertainties regarding the cost, public acceptability and secondary environmental impact of these processes, which require substantial enhancements to current nuclear infrastructures and capabilities.

(v)   There is a need to view waste management in a wider societal context.

There is, internationally, a heightened awareness of the wider context of waste disposal.[20] Discussions in and around national and international bodies make clear that attention should also be given to issues such as sustainability, which is a much larger societal question that requires a comprehensive look at energy production and use, as well as at waste disposal and treatment. Some of the points raised are noted here.

Box 2. **Waste management options**

Disposal:

For long-lived waste, the waste management community has developed the concept of deep geologic disposal in a repository that is both safe and resistant to malicious intervention. Long-term safety is based on a passive system of multiple engineered and natural barriers with a range of safety functions. This disposal concept does not preclude monitoring, maintenance and reversibility /retrieval, but these are principally measured to enhance confidence and should not, ideally, be required to ensure safety. Similarly, society may choose to use long-term institutional controls as a management tool but, even if such tools were to fail, human health and the natural environment should still be protected.

Disposal thus represents the radioactive waste management end-point providing security and safety in a manner that does not require monitoring, maintenance and institutional controls.

Extended surface storage:

In virtually all countries, some period of interim surface storage to allow decay of radiation and heat generation has always been recognised to be necessary or valuable. This interim storage is often at a centralised location, but can also be at individual facilities. The choice of location may reflect political- and social-acceptance problems more than technical or economic developments. Storage differs from disposal in that long-term monitoring, maintenance and institutional controls are necessary to maintain the safety and security of a storage facility.

More extended storage has been accepted as unavoidable because of delays in implementing disposal for both HLW and spent nuclear fuel. There are, in addition, arguments that favour extended (though not "indefinite") surface storage:

(i) Postponement of disposal is advocated by some scientists and decision makers who believe that more time is needed to prove the geologic disposal concept more completely and/or to allow public confidence to increase. Sweden is, for example, explicitly looking at the "zero option" of continued surface storage (in CLAB), in order to allow a broad based comparison of options for the next decades, in response to both public interest and Swedish legislative requirements.

---

20. This point was not reflected prominently in the responses to the NEA questionnaire – probably due to the limited technical role required of the respondents, who represented regulatory and implementing parties with specific mandates regarding radioactive waste disposal.

(ii) Electrical utility waste producers are under increasing economic pressures, due to the opening of the electricity supply market. Storage may be attractive for some utilities, largely because it can postpone large investments in expensive repositories. Whether the storage route is more economic on the intermediate timescale depends also on financial considerations such as whether utilities are any way required to establish segregated funds for future disposal facilities.

(iii) Increased reserves of uranium ores and decreased demand for recycled fissile materials have tended to make direct disposal of spent fuel more attractive. However, the fuel still represents an energy resource and, particularly in an age of sustainability, it can be argued that long-term storage keeps this resource easily available. Of course, other arguments concerning non-proliferation of fissile materials can lead to the conclusion that this availability is not a positive feature.

Indefinite storage, though advocated by some outside the waste management programmes, is not considered to represent a real alternative to deep geologic disposal. Continued surface storage appears to be rationally viewed not as a solution, but only as a postponement of disposal.

Partitioning and transmutation:

An approach which has been claimed to have the potential to change the future of geologic disposal is partitioning and transmutation (P&T) of long-lived radionuclides to give wastes which have shorter half-lives and therefore do not present as serious a challenge to the isolation capacity of repositories. P&T is being actively studied in Japan, Spain, France and the USA and (in a basic research-oriented way) in Belgium, Germany and Sweden. P&T approaches are in the early stages of development. Most of the work is analytical and prototype research with the intent to evaluate the merit of further development. P&T may be an option for well characterised, concentrated wastes, but is unlikely to be a practical option for the treatment of heterogeneous wastes with dispersed contamination. Pressure to devote efforts to this subject often appears to come down from government or advisory levels rather than up from the technical radioactive waste management community. The most pronounced expressions of technical attitudes towards P&T come from the US and the EC, which do not view the technology as a replacement for geologic disposal. At best, it provides a disposal inventory with less long-lived radionuclide content.

Storage and P&T are viewed by many in the waste management community as contributing to the process of radioactive waste management, rather than as its end point.

Consideration should be given to the possible adverse environmental impact of the exploitation of natural resources and, particularly, of energy sources, with emphasis not only on long-term protection of the environment, but also on sustainable development. Discussions within the scope of preparations for the most recent Collective Opinion indicated, however, that, in a waste disposal context, there is no proper consensus on the definition of sustainability or on methods to work towards this.

The impact of the repository should be evaluated, not only in highly technical terms of health hazard to the human population, but also a wider context more accessible to the public. This can involve relatively minor additions, such as showing the impact relative to the natural fluxes of radioactivity in the environment. Broad issues such as the impact relative to other societal activities could, however, also be addressed. The concepts of fairly distributing the burdens of civilisation amongst current generations or across future ones (intra- or inter-generational equity) should be addressed for waste disposal, as was done in the preparatory work for the most recent NEA Collective Opinion.

The waste management community would like to place nuclear waste disposal in perspective with other practices (involving both radioactive and non-radioactive materials) that may impact the environment, including regulation and licensing. It is, nevertheless, acknowledged that it is impractical to apply the comprehensive measures proposed for disposal of fuel cycle wastes to the management of all long-lived wastes from other sources. This is the case, for example, of radioactive waste from mining and technologically concentrated Naturally Occurring Radioactive Materials (NORM).[21] There is, nevertheless, increasing recognition that radioactivity is part of the human habitat, and that it needs to be viewed globally.

There is increasing recognition of the necessity to consider also the financial pressures, that affect the whole nuclear fuel cycle (e.g. deregulation of the electricity market), and are resulting in the reorganisation of many waste management organisations and may tend to favour a delay in final disposal. Efforts in the past, e.g. by the NEA, have led to various reports on cost comparisons. These are always difficult to interpret, however, and generally do not set the specifics within a framework that is sufficiently wide, particularly in a time of electricity privatisation and general globalisation of industries. The linkage of repository planning to the fate of the nuclear power industry is not only a cost issue. Some opposition to repository projects certainly arises from groups which do not wish to see a perceived problem of nuclear power being solved. Independently of the future use of nuclear power in the world, however, there is a clear need for development of deep repositories facilities. The large quantities of wastes existing today in civilian and military programmes must be disposed of in a safe manner. Nevertheless, the debate on disposal is inextricably linked to the continuing discussion on future strategies for supply of energy from nuclear and other sources.

As time passes the wider, societal issues are likely to become increasingly critical, as recognised in the strategic document of the RWMC.[22]

---

21. Because this difference has long been recognised, it is scarcely alluded to in responses to the questionnaire, apart from the Canadian reply where it is explicitly acknowledged that different approaches must be used.

22. *Strategic Areas in Waste Management – The Viewpoint and Work Orientations of the NEA Radioactive Waste Management Committee*. OECD/NEA, Paris 1999.

The international contribution

The importance of international fora and co-operation

Certain future needs are common to waste management organisations internationally. Others are constrained by national, rather than international, considerations. These constraints include the national legal and regulatory framework, national cultural considerations, the available geological settings and inventories of radionuclides to be disposed. Needs include the development of procedures and methods and suitably trained staff to implement them. Although the procedures and methods adopted will be nation- or programme-specific, they will be influenced by developments elsewhere. The development of these procedures and methods, the training of staff and progress in repository development will proceed most effectively through the development of international contacts and through the exchange of ideas internationally. It is therefore to be expected that international fora will continue to be important in meeting the future needs of waste management organisations. The fora provided, for example, by the NEA, provide a valuable mechanism for dialogue across several boundaries and for common projects, allowing the waste management community to be informed in a timely manner and at the proper technical depth on developments in all Member countries. The Collective Opinions organised and published by the NEA are the most extensively cited international documents which are judged to have made a positive contribution to the debate on deep disposal.

The importance of international co-operation is evident from answers to relevant questions in the NEA questionnaire. It is also evidenced by the impressive lists of successful co-operative projects, including rock-laboratory projects. Waste management organisations can gain technically from sharing insights from experience and even resources in co-operative underground-laboratory, natural-analogue and other technical studies. There has been a gradual building of confidence among, for example, OECD member nations in the scientific understanding of waste disposal processes, and in the work that is needed to produce credible evaluations of safety based on this enhanced scientific understanding. Co-operative efforts will also be helpful to legislators and regulators to:

- demonstrate the wide consensus that exists at the technical level;

- optimise use of technical and financial resources;

- clarify understanding of key concepts in repository development (e.g. the meaning of the stepwise approach to repository development);

- ensure that the process of repository development is fair, and is perceived as being fair by those outside the waste management community;

- work towards harmonisation of regulatory requirements across different types of environmental risks;

- (at least) rationalise differences between national regulatory guidelines.

International fora allowing cross-party dialogue and co-operative projects are thus likely to continue to play an important role in the future for all those involved in waste management. International projects can be of direct interest to small countries (which can benefit from economies of scale), to countries which may lack the resources needed for a proper disposal project (e.g. some of the former East Block states), to those with difficult environmental situations (e.g. dense populations), or to those which might benefit from having particularly well suited environments.

The readiness of all respondents to share the results of the NEA questionnaire results, testifies to the importance attached to openness between national and international programmes themselves, and between national programmes and regulators.

Overall conclusions of the review

A number of broad conclusions can be drawn from this report, the document referred to and input made available for the present review:

- Deep geologic disposal concepts have made significant progress in the past ten years, most especially in the technical areas concerning the understanding, characterisation and quantitative modelling of the natural and engineered safety barrier systems.

- No radical changes in strategy or in applied methodologies have proven to be necessary; although refinements are still being made, deep geologic disposal is effectively a technology that is mature enough for deployment.

- In many programmes, more emphasis is being placed upon the contribution of the engineered barriers, but the natural or geologic barriers in a deep repository continue to play a crucial role in determining the achievable long-term safety.

- All national programmes continue to support deep geologic disposal as a necessary and a feasible technology, even though some countries wish to postpone implementation of repositories or to evaluate other options in parallel.

- There is a general common trend towards advocacy of prudent, stepwise approaches at the implementational and regulatory level to allow smaller incremental steps in the societal decision making process. Discrete, easily overviewed steps facilitate the traceability of decisions, allow feedback from the public and/or their representatives, promote the strengthening of public and political confidence in the safety of a facility along with trust in the competence of the regulators and implementers of disposal projects.

- Although one deep geologic repository, purpose built for long-lived waste, is now operating, the timescales envisioned ten years ago for the development of deep geologic repositories were too optimistic. The delays that have occurred are partly due to operational causes, but mainly reflect institutional reasons, in large part associated with insufficient public confidence.

- There is an acute awareness in the waste management community of this lack of public confidence; efforts are needed by both implementers and regulators to communicate effectively to decision makers and the public their consensus view that safe disposal can be achieved.

- The implementers and regulators are more willing than ever to heed the wishes of the public in so far as these do not compromise the safety of disposal facilities. One common goal is to establish strategies and associated procedures that allow long-term monitoring, with the possibility of reversibility and retrievability. A number of programmes now consider these issues explicitly.

- Alternative means of radioactive waste disposal have often appeared to have promise prior to a thorough consideration of all aspects of the proposal. Several exotic options were studied earlier, and are no longer seriously considered. There are those who, for a variety of reasons, strongly advocate extended surface storage or partitioning and transmutation. The waste management community does not, however, regard extended or "indefinite" surface storage as a real alternative to geologic disposal; at best it offers a postponement of final disposal. Partitioning and transmutation is also not regarded as an alternative; at best it reduces the volume, or changes the isotope distribution, of wastes requiring deep disposal.

In summarising the status of the concept of deep geological disposal, it can be clearly stated that real progress has been made, and is being made, thanks to the extensive efforts of numerous experts working in diverse disciplines within national and international waste management programmes. Technical advances and improved societal interactions have taken longer than had been hoped and lengthy delays have occurred in the implementation of deep repositories. One disposal facility has, however, begun operation and a few deep disposal facilities are nearing the point when they will commence operation, although most such facilities are still many years away from implementation. The path towards their eventual implementation may well be eased by the marked increase in public confidence that is to be expected when the first deep repositories are in successful operation.

Appendix 1

THE NEA QUESTIONNAIRE

Table 1. **Respondents to the Questionnaire circulated by the RWMC of the OECD/NEA**

COUNTRY	ORGANISATION(S)
Belgium	SCK, ONDRAF
Canada	AECB, AECL, Ontario Hydro NRCan
Czech Republic	RAWRA
Finland	POSIVA, STUK, Ministry of Industry & Trade
France	DSIN, ANDRA
Germany	BfS
Hungary	HAEA
Japan	JNC
Korea	KAERI
Netherlands	Ministry of Housing & the Environment
Norway	IFE
Spain	CSN, ENRESA
Sweden	SKI, SKB
Switzerland	NAGRA, HSK
UK	EA, NIREX
USA	DOE-YMP, DOE-WIPP USNRC
EC	DGXI; DGXII
IAEA	Waste Safety Section

QUESTIONNAIRE

Please respond to all questions that apply to your organisation – where appropriate with the comment, "none" or "not applicable". Brief replies are sufficient, with references where appropriate. A few sections of the questionnaire require more detailed input (e.g. the enquiries on site characterisation and performance assessment issues). We encourage you to pass those sections on to other persons, if necessary.

1. National policies and strategies for long-term management of long-lived radioactive waste

1.1 Has there been any relevant new legislation or policy statements? (Please give references and brief highlights; note that new regulatory guidance is covered in Question 6).

1.2 What changes have occurred in organisational structures or in allocation of responsibilities?
   (a) for policy?
   (b) for implementation?
   (c) for regulation?
   (d) for financing?
   (e) other?

1.3 What changes have occurred in policy or waste management strategy?
   (f) overall strategy, e.g. relative weightings of disposal vs. storage, use of partition and transmutation?
   (g) time scales for implementation of stages of the strategy and facilities?
   (h) position on monitoring and retrievability?
   (i) co-operation within multi national programmes?
   (j) involvement with international organisations?
   (k) other?

2. Progress towards implementation of geologic disposal and related facilities

2.1 (a) Have their been any changes in technical strategy, major components or time scales of the programme for implementation of geologic disposal? and
   (b) Has your organisation made any official statements (or periodic statements) on the technical safety of underground disposal? (please give references)

2.2 What progress has been made towards development of geologic disposal and related facilities? For example, which facilities have:
   (a) commenced operation?
   (b) been constructed or are under construction?
   (c) been licensed for construction?
   (d) been the subject of a license application?

2.3   What has been the experience of siting of the above facilities, for example, with regard to:
     (a) selection of candidate and preferred sites (including selection criteria)?
     (b) legal or regulatory acceptance or rejection of chosen sites? (including reasons and lessons learnt)?
     (c) reaction of the public (especially locally; including the effectiveness of outreach initiatives)?
     (d) to what extent, and at what stages, has the siting process been influenced by geological criteria (e.g. general geology, specific site characteristics, ability to characterise geological environments or a specific site)?
     (e) what are the overall lessons learnt?

3.   Technical advances in waste conditioning and engineering aspects

3.1   Have significant advances been made in the techniques for geologic disposal in the following areas. (Distinguish between waste types if necessary).
     (a) waste conditioning?
     (b) packaging, encapsulation?
     (c) buffer materials?
     (d) other engineered barriers?
     (e) underground excavation and engineering?
     (f) other?

3.2   (a) Which of the above areas are priorities (and being actively worked on) in your national programme? *and*
     (b) What are the key remaining problems that need to be tackled?

4.   Technical advances in the practice of site characterisation

4.1   What regional or site characterisation campaigns have been carried out by your organisation since 1990? (Please give references, brief statement of purpose, whether results published etc.)

4.2   What information was sought, e.g. with respect to role of the geosphere in providing safe disposal, feasibility and constraints on excavation/construction, locating the repository, optimisation of design?

4.3   What significant technical advances have been made in:
     (a) site characterisation strategies?
     (b) field measurement techniques?
     (c) interpretation techniques?
     (d) other?

4.4   (a) Which of the above areas are priorities (and being actively worked on) in your national programme? *and*
     (b) What are the key remaining problems that need to be tackled?

5. **Practice of performance and safety assessment**

5.1 What performance or safety assessments have been carried out by your organisation since 1990? (Please give references, brief statement of purpose, whether results published etc.)

5.2 What significant advances have been made in evaluation of the *engineered barrier system* (EBS)? Please answer in respect of:
(a) characterisation and physical understanding, e.g. of waste form, EBS material behaviour, coupled processes?
(b) conceptual and mathematical modelling (detailed and assessment level)?
(c) overall confidence in EBS performance (identify key factors or arguments)?
(d) other?

5.3 What significant advances have been made in evaluation of the *natural barrier system* (geosphere and biosphere)? Please answer in respect of:
(a) characterisation and physical understanding, e.g. of geosphere, biosphere, and related processes?
(b) conceptual and mathematical modelling (detailed and assessment level)?
(c) overall confidence in geosphere and/or biosphere performance (identify key factors or arguments)?
(d) other?

5.4 What significant advances have been made in total system analysis, methodological and presentational techniques, and overall confidence of assessment? Please answer in respect any or all of the following:
(a) total system analysis? scenario development? treatment of uncertainty?
(b) quality assurance, verification, model validation?
(c) use of natural analogues?
(d) presentation of results?
(e) overall confidence, multiple lines of argument etc.?
(f) other?

6. **Underground rock laboratory programmes**

6.1 Do you have an underground research laboratory (URL) programme? If yes, please give references to work over the last decade and future programme if available.

6.2 Is your organisation directly involved in multi-national URL programme(s) or URL experiments in other countries? Please give references.

6.3 (a) What do you perceive as the key role(s) of underground rock laboratories within a disposal programme? *and*
(b) What specific benefits has your organisation gained from your own, or participation in other, URL programme(s) in the last decade?
(c) What do you see as key areas for investigation in future?

7. **Natural analogue programmes**

7.1 Do you have a natural analogue programme? If yes, please give references to work over the last decade and future programme if available.

7.2   Is your organisation directly involved in multi-national natural analogue programme(s) or natural analogue experiments in other countries? Please give references.

7.3   (a) What do you perceive as the key role(s) of natural analogue within a disposal programme?
        and

        (b) What specific benefits has your organisation gained from your own, or participation in other, natural analogue programme(s) in the last decade?
        (c) What do you see as key areas for investigation in future?

8.   Regulatory development

8.1   Have there been changes in regulations and/or guidance? Please give references to new regulatory documents and brief highlights of changes; mention draft and consultation documents if regulations/guidance not yet finalised. Specifically mention any changes in:
        (a) criteria, end points or their interpretation (e.g. dose/risk, other global or subsystem criteria, aggregation/disaggregation, timescales, non-radiological criteria)?
        (b) treatment of specific topics (e.g. human intrusion)?
        (c) coherence with conventional environmental legislation and planning law?
        (d) coherence with international guidelines?

8.2   Have there been changes in regulatory approaches? For example, with respect to:
        (a) the regulatory processes (e.g. phased approaches, interactions with implementers)?
        (b) consultation and communication (e.g. with stakeholders or public)?
        (c) integration with environmental and planning law processes
        (d) other?

9.   Trends in costs / financing / budgets

9.1   How much money has your organisation spent in the last ten years (or periods as convenient) on projects related to geologic disposal? If possible, give breakdowns against R&D vs. implementation/site development, and against HLW/SF vs. TRU/ILW.

9.2   What are the current trends (and forecast into the next few years) in size of your staff and costs allocated to projects related to geologic disposal? Give breakdowns as above if possible or in time.

10.   Public communication / involvement

10.1   Is there a formalised process for public participation in decision making related to waste management in your country? Who takes responsibility for this?

10.2   Do you see it as the role of your organisation to communicate (and to what extent) with any of the following?
        (a) general public?
        (b) political decision makers?
        (c) local affected public, or their representatives?
        (d) wider scientific and technical audiences?

and

Do you have specialist staff (how many) for any of the above purposes?

10.3   What initiatives has your organisation taken to communicate to the above audiences? For example by:
(a)   publications aimed specifically at these audiences?
(b)   presentations, exhibitions, meetings involving proponents and opponents, etc.?
(c)   other means?

10.4   What do you see as key information, or arguments, to transmit, e.g. technical safety of underground disposal, pragmatic need for disposal, quality of science, low risks, local benefits, other?

10.5   (a)   Have attitudes changed in your country during the last decade (general public, local communities or specific audiences)?
(b)   Which events have had major public impacts (positive or negative, and for what audience)?
and

(c)   To what extent has your communication and consultation had significant influences on either (i) attitudes in target audiences or (ii) your own programmes and proposals?

11.   Overall judgement

(including your own national programme, other programmes or international issues)

11.1   Name the 1-5 most significant developments in progress towards geologic disposal of the past 10 years?

11.2   Name the 1-5 most significant negative factors against progress towards geologic disposal of the past 10 years?

11.3   Name the 1-5 most significant developments required to improve progress towards geologic disposal over the next 10 years?

Appendix 2

OVERVIEW OF RESPONSES: STRUCTURAL AND ORGANISATIONAL, LEGAL AND REGULATORY DEVELOPMENTS

A2.1   Development of programmes, strategies

Wastes to be disposed

Relative to the situation 10 years ago, there appears to be an increased appreciation that geologic disposal is necessary, and not just for the vitrified HLW or spent nuclear fuel for which the first geologic disposal concepts were developed. Many countries now explicitly require repositories which will accept both spent fuel and HLW as waste forms. Examples are the USA, Switzerland, Germany, Belgium and the Czech Republic. ANDRA in France has also recently studied direct disposal of spent fuel, although the French National Evaluation Committee (CNE) has expressed reservations on this back-end strategy for spent fuel[1] (CNE 1998). Further, deep disposal is not required only because of the wastes arising from the commercial nuclear power fuel cycle; more diverse waste types are also now considered for geologic repositories. The WIPP facility in New Mexico is designed for transuranic waste (TRU) from the US defence programme. The geologic disposal of TRU waste from the nuclear fuel cycle has been evaluated in Japan and in several other countries. The inventories of low- and intermediate-level radioactive wastes (L/ILW) for deep disposal in the UK programme, and of long-lived wastes from reprocessing and other sources in the Swedish and Swiss programmes, are better quantified. "Historical" wastes produced by various activities in the past (military or industrial) are also being given increased attention. In the USA, a decision was taken in the 1980s to commingle the defence and civilian high-level wastes and spent nuclear fuels in the same repository. Fissile materials from the scale-back of the weapons programme are also to be disposed, but only after either a stabilising treatment, or as a spent fuel after conversion into and use as a mixed-oxide (MOX) fuel in a reactor.

New facilities

Within the last 10 years, new facilities for above-ground interim storage of spent fuel or HLW have been or are being implemented in various countries, such as Belgium, The Netherlands, Finland, Germany, Japan and Switzerland. Operating repositories are less common: in Finland, cavern disposal for LLW/ILW has been implemented at two locations along the lines of the Swedish SFR facility; while in Germany, operation of the Morsleben repository has been interrupted and Konrad awaits its final license,[2] in the USA, the WIPP facility for long-lived waste was certified for compliance with federal regulations, and now is in the process of obtaining certification of compliance with regulations administered at the state level.

---

1.   This is because spent fuel has residual value and, legally, may not be considered as an "ultimate" waste.

2.   The license of the Konrad facility is not expected in the immediate future.

The greater success in implementing interim storage may reflect the urgent short-term needs of the generators; the existence of increased surface-storage capacity tends to reduce the time pressure on implementing deep repositories for permanent disposal.

Self-sufficiency in disposal

During the last 10 years, there have been contradictory trends in the area of national self-sufficiency. Some countries such as Sweden, Finland or France, prohibit import or export of radioactive wastes for disposal; some, like Switzerland would allow export as an exception; some, such as the Netherlands, would welcome a co-operative repository solution. The potential for regional or international repositories, which would accept waste from more than one country. This is covered in an EU Community Strategy document, in a recent statement by the Council of Europe (1998), and in a recent IAEA report published in 1998 (IAEA, 1998).

Virtually all waste management programmes feel that they have a responsibility to propose national solutions to their own disposal problems. This appears to be mostly a reflection of current political and social acceptance realities, rather than a fundamental principle. Indeed, it is obvious that there are strong economic arguments to avoid duplication of effort, and achieving economies of scale.

A2.2   Legislation / Regulation

Developments in internationally recognised safety standards

The International Commission for Radiological Protection (ICRP) formulates fundamental radiation principles and criteria for world-wide application. Ten years ago, the most recent advice on waste management from the ICRP was its Publication 46 in 1986. More recently, Publication 64, published in 1993, established radiation protection principles for a number of practical situations where exposure is not foreseen, but at the same time cannot be excluded. The ICRP has elaborated its general guidance on policy and ethical considerations relevant to radioactive waste management in ICRP Publication 77, published in 1998. A report on "Radiological Protection Principles for the Disposal of Long-lived Solid Radioactive Waste", is in an advanced stage of preparation. It will address potential exposures from long-lived wastes, protection objectives in the long term, the weight to be given to future doses and the application of the concept of "optimisation of protection".

In its reports, the ICRP have included advice that is specifically relevant to the performance assessment of repositories for long-lived waste. For example, Publication 64 comments on the problems associated with a changing biosphere and gives guidance regarding scenario selection, offering a rationale for the exclusion of particular scenarios in a safety case presented as part of a license application. Furthermore, Publication 77 makes the practical suggestion that, in order to ensure adherence to the principle that future generations should enjoy the same protection as the present, the annual individual effective dose to a critical group for normal exposure and the annual individual risk to a critical group for potential exposure will together provide an adequate input to a comparison of the limiting detriment to future generations with that which is currently applied to the present generation." (ICRP 77, paragraph 69).

Based on the work of the ICRP, the International Atomic Energy Agency (IAEA) has issued internationally recognised non-binding standards on radioactive waste safety, the Radioactive Waste

Safety Standards (RADWASS). These are intended to establish an ordered structure for safety documents on waste management and to ensure comprehensive coverage of all relevant subject areas.[3] In the area of deep geologic disposal, the most recent safety guidance from the IAEA is contained in Safety Series No. 99, published in 1989. The IAEA has, however, established a working group to explore and, where possible, develop consensus on relevant issues[4] and the findings, when completed, together with guidance from the ICRP, are likely to be used in developing new safety standards in several nations.

Agreements among states

Over the past decade, several agreements among sovereign States have been implemented with the support of the IAEA. These agreements may be viewed as a component of a global framework for fostering intergovernmental collaborative efforts in the area of nuclear safety. Specifically regarding radioactive waste disposal, the "Joint Convention on the Safety of Spent Fuel Management and of the Safety on Radioactive Waste Management" was adopted in 1997 and, when ratified, will be the first legally binding instrument in this area. The technical content of the Convention is based on a number of ethical principles related to radioactive waste disposal, as well as safety principles and concepts, that are contained in the IAEA document "The Principles of Radioactive Waste Management", published in 1995. The implementation of the Convention will formally be pursued through peer review of national reports at Review Meetings of the Contracting Parties.

Trends in national laws and regulations

At the level of environmental or nuclear law, various countries have undergone significant developments during the last 10 years. Most active have been legislators in the USA, where there is the most prominent tendency for politicians and lawmakers to take direct actions to move waste disposal projects in a desired direction. However, new environmental or nuclear laws have been passed also in the UK, Germany, the Netherlands, Hungary and the Czech Republic. The Canadian parliament in 1997 passed laws streamlining and reorganising its regulatory structure and adding further assurance that license applicants have the financial wherewithal to pay for the disposal of their nuclear wastes. At the regulatory level there has also been activity in countries like the USA, Switzerland, Sweden and Finland, all of which have amended, or are amending, their regulations. Some countries, like Belgium and The Netherlands, do not, as yet, have regulations specifically governing radioactive waste disposal, and some like Spain and Japan are in the process of developing such regulations.

There are some general trends worthy of discussion. Issues of interest to many legislators and regulators include the allocation of financial responsibilities for the back-end of the fuel cycle, the definition of long-term safety goals and the formulation of appropriate compliance criteria for meeting these goals. One feature, noticeable by its absence, is the embedding of laws or regulations for waste disposal in a wider environmental regulatory framework. Although most countries would claim

---

3. The basic principles and concepts of safe radioactive-waste management are set out in Safety Series No. 111-F, published in 1995.

4. The IAEA's Working Group on Principles and Criteria for Radioactive Waste Disposal has issued three reports on safety-related issues: *Safety Indicators in Different Time Frames for the Safety Assessment of Underground Radioactive Waste Repositories*, TECDOC-767, *Issues in Radioactive Waste Disposal*, TECDOC-909, and *Regulatory Decision Making in the Presence of Uncertainty in the Context of the Disposal of Long Lived Radioactive Wastes*, TECDOC-975.

coherence with other environmental legislation, there is little direct mention of issues like sustainability or broad-based approaches at the legislative level to assure a uniform degree of environmental protection from a variety of radioactive materials and from non-radioactive hazardous waste forms. An exception is the UK, where the Environment Agency is required to address sustainable development under the provisions of the Environment Act 1995. This would include addressing sustainable development in relation to radioactive waste disposal.

Where amendments to regulations have occurred, they are often in the area of criteria for long-term safety. Common specific trends are:

- a greater use of risk-based criteria;

- acceptance that safety analyses must consider very long times (10 000, 100 000, and more years);

- recognition that rigorous compliance cannot be demonstrated, especially at timescales beyond 10 000 years;

- appreciation of the potential value of safety indicators other than dose or risk;

- recognition that scenarios involving human intrusion into a sealed repository require separate consideration.

A more general common trend is towards advocacy for stepwise approaches at the regulatory level to assure smaller steps in the societal decision making process. Clear definitions of what this means in practice are lacking, but specific examples suggest there is more chance of successful licensing if applications are made for discrete, easily overviewed steps in the development. One example is in Switzerland, where a prime reason for loss of a public referendum at Wellenberg (Kowalski and Fritschi 1996) was that the public preferred to give an initial permit for the construction of an exploratory tunnel rather than directly for the final repository. The complexity of the public debate, on the other hand, is illustrated by the UK counter example, in which permission for an exploratory facility at Sellafield was refused (CUM 1997), even though it was made quite clear this facility would only be used for exploration to support a future finding on the suitability of the proposed location.

A2.3   Organisation of implementing and regulatory bodies

There have been few fundamental changes in the organisation of the implementing side of waste management. Significant is the founding of further dedicated implementer organisations, e.g. in Finland (Posiva), Czech Republic (RAWRA), Hungary (PURAM) and Switzerland (GNW). In fact, the model of an implementing body driven primarily by the waste producing organisations is growing in popularity. Even in some countries (e.g. Japan, Germany, Canada) which do not currently have this type of organisation, consideration is being given to establishing one. In Canada, the role that the Atomic Energy of Canada Limited organisation played in the defining of a geologic waste disposal concept has ended, and the major nuclear utility, Ontario Hydro, has taken on more direct responsibility for furthering developments toward the disposal of its radioactive wastes.

At a different level in the implementation process, some countries have made organisational changes aimed at facilitating the contentious issue of siting. France introduced a mediator who had success leading to volunteer communities for underground laboratories. The USA, on the other hand, had little success with a negotiator for siting a final or an interim storage facility, and has dropped the idea.[5] Sweden has recently appointed a National Co-ordinator with a similar role in the siting process.

There has been a trend during the last ten years towards changes which stress the independence of nuclear regulators and nuclear implementing organisations. For example, responsibility for regulation in the Czech Republic transferred from the AEC (Atomic Energy Commission) to a new organisation SONS (State Office for Nuclear Safety); in the UK the EA (Environment Agency) has been formed; in France the ISPN is being separated from the CEA, and in Canada a new Nuclear Safety Commission is replacing the former Atomic Energy Control Board. Also within international organisations there have been changes aimed at clearly delineating the nuclear safety aspects of waste disposal from the technological aspects. In both the EC and the IAEA, regulatory matters have been organisationally split off from R&D activities on waste management. Further structural changes have been implemented within the nuclear regulatory organisations of, for example, Spain and Sweden – in each case leading to waste management issues having a higher profile within the government.

A characteristic feature of the NEA's work on radioactive waste management has been its support for the discussion of technical, policy, and strategic issues in fora which include both regulators, implementers, and policy makers. The sharing of insights into technological limitations, as well as advances, is important if regulations are to be implemented in the processes of critical reviews and hearings which determine if a facility is licensed to operate or not. In spite of this, the NEA has also seen the need to provide fora for regulators alone to share their insights and approaches across national boundaries. For example, the RWMC formed a sub-committee of regulators in 1998.

A2.4   Costs and financing

Budgets, costs

Achieving some degree of coherence in studies of costs in different programmes is difficult even in the scope of projects specially set up for this purpose (cf. the LLW and HLW cost studies of the NEA (NEA 1993 and 1999a); the DGXI study contract report on the subject, currently being published).

It is apparent that very large sums of money have been devoted to trying to develop deep repository projects. These run into billions of dollars in some cases, with the USA and Germany, both of which have identified single candidate sites, having invested the most. Even more modest programmes like those of Belgium, the UK, Canada and Switzerland have invested hundreds of millions of dollars without reaching a consensus on siting.

Mature programmes tend to show a trend towards becoming slower and smaller, although there is also increased emphasis on assuring long term funding, reflecting perhaps acceptance that the task is more long-term than believed earlier. An exception occurs when real progress is expected in the foreseeable future; for example, Sweden anticipates growth in budget of its HLW disposal programme as soon as it has can overcome its immediate siting hurdle. In the newer or less mature programmes,

---

5.   Significant interest for siting an interim storage facility was found. The US congress, however, stopped funding the efforts and the Office of the Negotiator expired. Sites identified through the Negotiators Office are currently being pursued as private storage facilities.

increasing efforts are being made to establish viable, long-term disposal projects. This is often done by collaborating with long-established programmes, within the scope of joint projects or bilateral agreements or else in a contracting relationship.

Funding of waste management

There is continuing growth in the application of the "polluter pays" principle to the funding of waste management. This principle was incorporated in Swedish legislation in the 1970s and, more recently, in the Swedish Financing Act of 1992 and in the financial reforms of 1996. The USA has long had a system in which fee is levied on nuclear electricity to fund the work done by the DOE on management of commercial wastes from reactors. Germany in its Atomic Energy Act of 1998 (GER 1998), Canada in its Nuclear Safety and Control Act of 1997, Hungary in its 1996 Act No. CXVI on Atomic Energy, and the Czech Republic with its Atomic Law of 1997 have all joined the extensive list of countries which already have the principle legally anchored. Also in other countries, such as Belgium and Switzerland, formal governmental measures have been taken to assure that sufficient funding is available to close the back-end of the fuel cycle.

Appendix 3

OVERVIEW OF RESPONSES: SCIENTIFIC AND TECHNICAL BASIS

In this Appendix, the responses to the questionnaire have been integrated to provide a review of the status of the scientific basis underlying waste disposal projects. The following Appendix 4 goes on to examine in more detail the specific challenge of performing credible safety analyses based on a proper understanding of the system behaviour. A generalised overview of the status of these two areas is feasible because there were no fundamental differences between opinions offered by the respondents. Although there were nuances in judging the completeness of the technology and safety methodology, there was broad consensus on all key points.

A3.1   Waste conditioning, repository engineering and design

Progress in the areas of waste conditioning, repository engineering and design has been achieved mainly by gradual improvements in conceptual design and in the technology that is available to implement them. There have been relatively few new design concepts or technological breakthroughs; the trend has been more towards incremental improvement and the more rigorous demonstration of existing concepts. Examples of significant technical changes over the last 10 years are:

- The decrease in use of bitumen as a conditioning material for nitrate-containing wastes and for TRU waste (in Belgium and by Cogema in France) and the greater emphasis on glass and ceramics.

- The extension of the use of vitrification technology to wastes other than HLW, such as mixed TRU wastes.

- The development of composite canisters for spent fuel (steel insert with outer copper sheath) in Sweden and Finland, for HLW (composite of titanium or copper with steel) in Japan, and for both spent fuel and HLW (the dual metal disposal container) in the USA. Some programmes, however, e.g. Switzerland, Japan and France, retain steel on its own as a reference material.

- Specially formulated porous cement-based backfills developed in Switzerland and the UK. Such backfills are intended to buffer inflowing groundwater to high pH, favouring the low solubility and high sorption of key radionuclides, while allowing the free escape of gas generated by the degradation of repository materials and wastes.

- The porous backfill and "drip shields", recently proposed as design options in the US, which are intended for thermal management and to protect the dual-metal waste packages from dripping water, thus extending their life.

- The development of computer technology, communication and control systems that will allow the use of remotely controlled equipment for waste-package emplacement and related operations.

In several countries, a range of possible container materials and designs have been or are still being considered, with associated research on material properties and long-term corrosion performance. However, a single reference design is often adopted in feasibility and safety assessments. There is increasing emphasis on demonstrating the feasibility of reliable fabrication and also on assuring long-term performance, and this has fed back into container design, e.g. in the case of development of the copper-steel canister mentioned above, or the consideration of Alloy 22 in the USA and the consideration of both copper and titanium in Canada. It is seen as imperative to demonstrate the feasibility of techniques to manufacture containers that meet pre-defined specifications (e.g. the assurance of uniform welding). This has been done, for example, in Sweden where full size copper canisters have been fabricated as prototypes and in the US, where full-diameter, quarter-length prototype containers of the current two-metal waste package design have been manufactured to demonstrate the feasibility of the "shrink-fit" process.

The emphasis on fabrication technology and on safety performance under repository conditions also applies to investigations on buffer/backfill materials, including the compatibility between different materials (e.g. cement/bentonite). In the case of bentonite, this emphasis is demonstrated in the FEBEX experiment (EC 1998e), which has examined the industrial fabrication, handling and emplacement of a bentonite backfill, and its behaviour under realistic repository conditions, and in international collaborative projects to investigate the possibilities for gas release through a bentonite layer. Understanding of the long-term chemical alteration of bentonite and cement, and the development and validation of models for these processes, is also seen as a priority area.

In some countries, approaches have broadened to cover disposal of HLW together with direct disposal of spent fuel (SF) and also to consider co-disposal of other long-lived waste streams. Thus, the range of different waste types considered has increased. In particular, the need to deal with historical waste arisings, that may have been in store for many years, and the waste from decommissioning programmes, is appreciated. In the UK, for example, there is now an active programme to recover, condition and encapsulate historical waste. Each waste type has its own associated challenges with respect to disposal technology. In the USA there is a massive programme aimed at clean up of DOE sites, with disposal of all resulting residues, some of which may need disposal in a HLW repository. In the USA there is also a need to treat and dispose of surplus weapon materials resulting from reducing the nuclear arsenal (DOE 1997).

It is obvious that the overall performance of the engineered barrier system can be influenced by the properties of the waste form. These properties may in turn be influenced through controlling the composition or the conditioning process. In the case of spent fuel, adding tailored fillers to provide more desirable performance characteristics may also be an option. Not every respondent indicated that conditioning or supplementing the waste form was part of the engineered system design strategy, but some noted that this was a potentially important part of creating an integrated natural and engineered barrier system.

The predominant strictly technical view is still that disposal implies emplacement without intention to retrieve, and that allowing for retrievability must not be allowed to compromise the long-term performance of a repository. Nevertheless, public and political concern has led to increased interest also in the technical community in the possibilities for including design measures or adapting operational plans to facilitate retrievability. In the UK and Switzerland, for example, the low strength of the proposed backfill material in L/ILW repositories is seen by the developer as advantageous, since it facilitates retrievability, should this be required. In several countries, e.g. France, Canada, Switzerland and the USA, the possibilities for, and technical implications of, keeping a repository open and monitored for an extended period following emplacement of wastes are being actively evaluated.

The key general conclusion that can be drawn concerning the development of disposal technology is that, by and large, the necessary technology is available and ready to be deployed when public and political conditions are favourable. This position is, in fact, directly articulated in the responses of Ontario Hydro in Canada, and is indirectly indicated in the observation by Canada's Nuclear Fuel Waste Management and Disposal Concept Panel which noted that technical safety may have been demonstrated, but that lack of public acceptance warranted a different approach be taken (CAN 1998).

A3.2   Site characterisation

Ten years ago, many national and international programmes perceived the lack of identified sites, and of site-specific data, to be a major limitation to the degree to which the adequacy of a repository design, with respect to long-term safety and construction feasibility, could be assessed. In the intervening period, considerable work has been carried out to rectify this deficiency. Examples are the extensive site characterisation work at Sellafield in the UK, Yucca Mountain and WIPP in the USA, Gorleben in Germany, Mol in Belgium, Wellenberg in Switzerland and at the potential deep disposal sites in France.

Two lessons can be drawn from recent experience in site characterisation work. One is that conflicting data can emerge, implying that yet more data may be required to understand the system – i.e. to reduce the uncertainty in a system that is more complex than originally believed. This happened, for example, at Yucca Mountain in the USA with ^{36}Cl measurements that suggested the existence of heretofore unmodelled faster flowpaths and led to a restructuring of the modelling of flow (Wolfsberg, A.V. *et al*). The second lesson is that even with extensive data bases and favourable performance calculation results, success in a permitting process is not guaranteed. Technical disagreements between proponents and opponents can stall programmes (e.g. at Sellafield in the UK); the political process may cause blockages (e.g. at Konrad in Germany or Wellenberg in Switzerland); or the lack of public confidence may require other steps to be taken before progress can again be made (e.g. the Canadian experience).

A more integrated strategy of site characterisation has been developed in several national programmes. For example, probabilistic techniques to optimise the site characterisation strategy have been developed, and the USA, as well as others, have performed formal system evaluations to focus experimental and design work on areas that promise to improve system safety and confidence. The need for an interactive project management structure has been recognised. This should enable the effective co-ordination of the planning and implementation of site investigations, the project-specific evaluation of site properties, the development of performance assessment methods and the application of these methods in comprehensive assessments.

Recent emphasis in site characterisation has been on the needs of safety assessment and, in particular, on the understanding of water conducting features and the flow of groundwater through them. This is generally recognised as the most likely means by which any radionuclides released from a repository might be conveyed to the surface environment. Other aspects of characterisation that are important to safety assessment, such as hydrogeochemical analysis and the analysis of the roles of colloids, organics and potential microbial processes, have also received attention. Most waste streams, however, do not contain any significant levels of potentially important organics. Furthermore, the

ARCHIMEDE project (Griffault, L. *et al* 1996) in the Boom Clay in Belgium suggested that, in that particular type of host rock, high natural organic content played no significant role in radionuclide transport.

Heterogeneity has become increasingly recognised as a universally present, and highly safety-relevant, feature of the geological environment. Even in relatively uniform rock formations, such as the Boom Clay in Belgium, lithological and tectonic heterogeneity may need to be characterised in order to provide an adequate geological database for performance assessment.

The emphasis on the acquisition and interpretation of hydrogeological data has been noted in the EC Mirage and THM workshops (EC 1995 & 1995a) and in the NEA GEOTRAP workshop series (NEA 1997a, 1998, and 1999c). In some programmes, less attention has, up to now, been paid to aspects of site characterisation that are required in order to assess engineering feasibility, to optimise repository design and to judge whether the risk of inadvertent human intrusion is acceptable (e.g. physical and geotechnical properties and the presence of natural resources that might attract future exploration of a site). Advances have, however, been made in the specification of specific technical criteria for the acceptability of a potential host rock (including, in the case of Sweden, acceptance criteria for individual deposition holes (SKB 1998, 1998a). Such criteria, if developed in advance of an exploration programme, not only serve to guide that programme; they may also raise the confidence of the public that an implementer would be prepared to abandon a site, should it prove unacceptable. For this latter reason, the criteria should not be unrealistically stringent or they may lead to the abandonment of a site which is adequate in terms of its ability to assure public health and safety.

Techniques for measurement and interpretation of data have continued to be refined and tested at potential sites and at underground research laboratories. Advances have been made, for example, in:

- detection of small groundwater flows in deep boreholes;

- extraction of undisturbed groundwater samples from low-permeability media;

- in the case of unsaturated media, the use of environmental isotope sampling methods to identify pathways for past and potential future rapid recharge of water;

- determining by electromagnetic methods the depth of more saline waters.

The interpretation of data is, in particular, viewed by most programmes as a priority area. There has been increased use of:

- three-dimensional CAD/visualisation techniques as an aid to the interpretation and integration of interdisciplinary data [e.g. geological, hydrogeological and geochemical data in the SKI SITE 94 performance assessment (SKI 1996)];

- numerical modelling tools for the representation of heterogeneity, including probabilistic modelling tools to take account of incompletely characterised natural variability (in virtually all programmes);

- the interpretation of hydrochemical data to indicate patterns of groundwater flow.

Furthermore, the performance of processing and interpretation codes has improved, such that field data (e.g. from seismic surveys) are able to yield more relevant information than previously.

The further development of measurement techniques that do not perturb the characteristics of a rock that must remain intact to provide safety is also seen as important (e.g. EMR techniques). For countries approaching a systematic siting process, such as Canada, electronic tools for geologic mapping and methods for integrating data from remote sensing and land-based sources are priorities. Other remaining problem areas include:

- bounding the effects of infrequent, but highly transmissive pathways for radionuclide transport;

- determining infiltration and groundwater recharge (in the case of the Yucca Mountain Project in the USA);

- the influence of gas on the barrier properties of the host formation;

- the characterisation of naturally occurring colloids;

- the influence of organic matter as complexants for migrating radionuclides;

- natural and induced changes to the geosphere.

Overall, experience over the past ten years has shown that, while some transfer of geological experience between sites is possible, issues are often site-specific and site-specific solutions are therefore often necessary. In general, it is recognised that high confidence in the proper safety functioning of the geologic barrier is harder to achieve than was thought 10 years ago. This explains to some extent the tendency, in many programmes, to place increasing emphasis on the role of the engineered barriers in the disposal system.

It is important to place this observation in perspective, however. Prior to site characterisation, modelling of site performance may, of necessity, be based on a rather idealised view of natural system properties, and may not reflect the realities and uncertainties understood after characterisation. Typically, this means that when more is known, more is also known about heterogeneity and uncertainty. A robust engineered system can help to counter this uncertainty, occasionally at the expense of producing a rather over designed system. The natural system must, at a minimum, ensure that suitable near field conditions are achieved in the repository. This means it must protect the engineered system from human intrusion and from the rapid changes experienced at and near the ground surface; it should also provide an environment favourable to the longevity of the materials of which the engineered system is made (in terms, for example, of geochemistry and groundwater flow). In every case, the vast majority of radioactivity in the emplaced inventory, in a practical sense, never leaves the near field region where it is emplaced. It is only a minor part of the radionuclide inventory that may potentially be released from the engineered system, and this is transported away from the repository only slowly along tortuous pathways, or diluted through mixing with uncontaminated waters. Thus, the current emphasis on solid assurance through engineering should not detract from the fact that the geologic setting of the repository remains a vital part of the overall deep disposal system. Furthermore, the uncertainties associated with engineered systems should not be overlooked.

A3.3   Use of underground rock laboratories

Underground rock laboratories (URLs) fall into two broad categories:

- "Generic" URLs, used principally to obtain information on a broad rock type, to develop understanding and test models and to develop experience in techniques relevant to site characterisation and to repository construction and operation. Examples are Asse in Germany, Stripa and Äspö in Sweden, Grimsel and Mt Terri in Switzerland, URL in Canada, Tono in Japan, and Tournemire in France.

- Site-specific URLs that form an integral part of the investigation and development of a potential site, before a repository is built. Examples are WIPP and the Yucca Mountain Exploratory Studies Facility (ESF) in the USA, HADES/URF at Mol in Belgium, the planned French labs, and the RCF sought by Nirex in England.

In either case, the costs involved in the development and operation of a URL, and the possibility of sharing existing knowledge and experience, can make international co-operation in underground studies advantageous, as is the case for most of the facilities in the list of URLs in Table A3.1. Further generic and site-specific URLs are currently planned, such as those at Meuse in France, at Pribram in the Czech Republic and Horonobe in Japan.

Table A3.1. **Underground rock laboratories and other underground research centres [based, in part, on Kickmaier & McKinley (1997)]**

Generic URLs			
Rock laboratory	**"Host rock" (depth)**	**Organisation**	**Remarks**
Forschungsbergwerk Asse, Germany	Salt dome	GSF	R&D programme closed down in 1995
URL Lac du Bonnet, Manitoba, Canada	Granite (240-420 m)	AECL	operating since 1984
Tono, Japan	Sediment	JNC	galleries in former uranium mine, operating since 1986
Kamaishi, Japan	Granite	JNC	galleries in former Fe-Cu mine, completed in 1998
Mizunami URL, Japan	Granite	JNC	borehole drilling underway
Stripa mine, Sweden	Granite (360-410 m)	SKB	galleries in former Iron mine, operated from 1976 to 1992
Äspö Hard Rock Laboratory, Sweden	Granite (< 460 m)	SKB	Construction started 1990
Grimsel Test Site, Switzerland	Granite (450 m)	NAGRA	gallery from a service tunnel of a hydroelectric scheme, operating since 1983
Mont. Terri Project, Switzerland;	Opalinus clay (hard clay 400 m)	SHGN	gallery from a highway tunnel, initiated 1995
Fanay-Augères, France	Granite	IPSN	galleries in uranium mine, operating from 1980-1990
Tournemire facility; France	Sediments (hard clay 250 m)	ANDRA, IPSN	former railway tunnel and adjacent galleries, operating since 1990

Table A3.1 (continued). **Underground rock laboratories and other underground research centres [based, in part, on Kickmaier & McKinley (1997)]**

SITE-SPECIFIC URLS			
Rock laboratory	**"Host rock" Depth**	**Organisation**	**Remarks**
High Activity Disposal Experiment Site (HADES), renamed to Underground Research Facility (URF) Mol/Dessel, Belgium	Boom clay (plastic clay, 230 m)	SCK/CEN,	Ondraf/Niras; 1980 (shaft sinking), operating since 1984 and extended 1998-1999
Olkiluoto research tunnel; Finland	Granite, (Tonalite 60-100 m)	Posiva	Olkiluoto repository for low- and intermediate-level waste operating since 1992
Gorleben, Lower Saxony, Germany;	Salt dome (> 900 m)	BfS, DBE	potential repository site, shafts constructed 1985-1990, galleries under construction since 1997
WIPP Carlsbad, New Mexico, USA	Salt (bedded), Salado Formation, (650 m)	USDOE-CAO	operating since 1982
Exploratory Studies Facility (ESF), Yucca Mountain, Nevada, USA	Calico Hills Welded tuff (300 m)	USDOE	in situ testing began in 1996; construction of an exploratory side tunnel completed in 1998

Note: Gorleben has been selected for underground investigations aimed at demonstrating site suitability.

There is consensus among most organisations involved in URL projects as to the information that can usefully be obtained from studies in these laboratories. This information includes the aspects listed in Table A3.2.

Table 3.2. **Information obtained from rock laboratories and other underground research centres [based, in part, on Kickmaier & McKinley (1997)]**

Class of information	Examples
Further development and testing of excavation techniques.	• URF: demonstration of technical feasibility of gallery drilling in plastic clays. • Olkiluoto: studies of the performance of disposal technologies.
Quantification of impact caused by excavation (regional and local scale; physical and chemical perturbations).	• ZEDEX experiment at Äspö. • EDZ experiment at GTS. • Experiments at Mont Terri. • Thermal-mechanical-hydraulic tests at URL, Canada.
Application of site-exploration strategies and strategies to adapt underground systems as more information is acquired.	• Full-scale deposition holes research tunnel to Olkiluoto. • Application of geophysical methods at GTS and Stripa.
Integration of results to derive conclusions, conceptual models and predictions regarding groundwater flow (and 2-phase flow).	• Task forces on groundwater flow and modelling at Äspö. • Experiments at Mont Terri. • Unsaturated zone seepage tests at ESF.
Testing of models, exploration methods and processes potentially relevant to radionuclide transport through rock.	• Radionuclide Retardation Project at GTS; • Experiments at Mont. Terri. • Unsaturated zone transport tests at ESF. • Solute transport and diffusion experiments at URL, Canada.
Simulation of effects caused by emplacement of radioactive waste (heat, nuclide release, mechanical impact).	• CERBERUS at URF. • TSS project at Asse; • FEBEX project at GTS; • Heater test at Stripa, ESF and GTS. • Planned experiments at Mont Terri. • Thermal-mechanical-hydraulic tests at URL, Canada.
Demonstration of engineered-barrier systems (feasibility).	• Borehole sealing and buffer mass tests at Stripa. • FEBEX project at GTS. • Planned experiments at Mont Terri. • Buffer and container testing at URL, Canada. • RESEAL project at URF.
Experiments related to long-term processes, post-operational phases, geochemical corrosion, geomechanical stability, etc.	• The PRACLAY concept demonstration in Belgium. • In-situ test on coupled thermo-hydraulic- mechanical processes and model validation and Kamaishi. • Planned experiments at Mont Terri. • Swedish demonstration repository. • Thermal-mechanical-hydraulic tests at URL, Canada.

The emphasis on different aspects can, however, differ between organisations. In particular, some issues are rock-type and disposal-concept specific (e.g. the use, or not, of cement as a construction material will determine the relevance of tests on the effects of a hyperalkaline plume in a repository in the unsaturated tuff of Yucca Mountain). Nevertheless, there is a tendency towards increased international collaboration, both at the level of devising rock laboratory R&D programmes (e.g. Äspö, GTS and Mont Terri) and in specific experimental studies (e.g. EU-sponsored studies at the URF in Belgium and Asse in Germany).

Some trends in R&D programmes at URLs can be identified (Kickmaier & McKinley (1997), complemented by questionnaire responses). There is a decreasing emphasis on basic feasibility studies and on the accumulation of fundamental geological data; increased effort is being focused on optimisation of methodology and on testing of key performance assessment models. This is true for the generic underground laboratories more so than for underground investigations of the suitability of a specific site, of course, where the fundamentals need to be established as the foundation for the modelling of system performance at that location. There is also increasing importance placed on full-scale "demonstration-type" experiments on engineered barrier systems for high-level waste and spent fuel in URLs, e.g. FEBEX at GTS (EC 1998e), RESEAL (EC 1998d) and PRACLAY (EUR 18047) at the URF in Belgium and the planned demonstration repository in Sweden. This may reflect the increased emphasis being placed on the engineered barrier systems in some performance assessments, as noted above.

An indirect benefit of the trend towards collaborative projects is the development of international and interdisciplinary contacts that may be valuable in other aspects of repository development. Finally, direct visits to URLs and reference to URLs in publicity material may contribute to public confidence in the feasibility of safe disposal.

A3.4   Use of natural analogues

Natural and anthropogenic analogues (and natural system studies that examine the transport of natural tracers in groundwater), and analogue studies at potential sites, are widely viewed as having the potential to complement the data obtained from site characterisation techniques and other experiment studies in underground rock laboratories (e.g. field tracer transport experiments). In particular, they may overcome the limitations of spatial and temporal scale that are inherent in such techniques (NEA 1997a). Furthermore, at the level of public information, they are potentially valuable in illustrating that a strategy of long-term geological isolation of waste is reasonable.

Their interpretation is, however, complicated by lack of information regarding, for example, the initial conditions of the analogue system (this is less a problem for anthropogenic analogues, although these analogues, of course, provide information for relatively short time scales). Data from natural analogues have thus rarely been used directly in repository performance assessment. Rather, natural analogues are seen as a component of the confidence building process. They support the qualitative and (less commonly) quantitative understanding of key processes. Some examples of this are:

- Insights on the oxidation of uranium and the migration of uranium dioxide through NRC studies of analogue sites in Mexico and Greece; (Smellie *et al.* 1997) (Murphy *et al.* 1997).

- The processes controlling the performance of spent fuel in a repository in plutonic rock, in a reducing environment and protected by clays are better understood from work in the Cigar Lake uranium deposit natural analogue in Canada (Cramer, J.J. and Smellie, J.A.T.).

- Studies of natural clays cited by Nagra provide evidence that, under repository conditions, the swelling capacity, permeability and cation exchange capacity of bentonite will be retained over a very long period (Miller *et al.* 1994, Chapter 4.4).

- Evidence for the stability of cement gels, for the absence of colloids and organic complexants and for very low levels of microbial activity at the Maqarin site, in Jordan and in Oman (EC 1996, SKB 1998b).

- The observation of rock matrix diffusion of uranium in granite at El Berrocal (EC 1997c), Spain in crystalline rocks in Northern Switzerland and at the Grimsel Test Site (Miller *et al.* 1994, Chapter 5.3).

- Validation by the USDOE of thermochemical data used with the EQ3/6 code to model rock-water interactions, through studies of geochemical data from wells at the Wairakei geothermal field in New Zealand; (Glassley 1994).

- Pocos de Caldas, a major international study looking at solubility and transport of radionuclides in and around natural uranium and thorium ore bodies (Miller *et al.* 1994).

Natural analogues also provide evidence that no potentially significant long-term processes or phenomena have been overlooked (NEA, 1997) and may be used to gain general confidence in the behaviour of geological aspects of a repository. More quantitative comparisons of model predictions with observations from analogues, and the better integration of natural analogues into performance assessments, are seen as areas warranting further development.

Natural analogue exercises are often performed co-operatively, with several organisations involved in each project. Examples are the NEA ARAP and ASARR projects, which focused on the Koongarra uranium ore deposit in the Northern Territory of Australia (ARAP, 1992; ASARR, 1996) and the work of the EC Natural Analogue Working Group (EC, 1997a; EC, 1998). As a result, participating organisations have benefited through the development of international and interdisciplinary contacts.

Current Canadian work includes plans for a new analogue addressing copper corrosion properties at a Newfoundland site, and the completion of ongoing work addressing bentonite properties at a deposit of that material in Saskatchewan. In the USA, there are plans taking shape to be able to provide supporting data from natural analogues during the licensing phase and after. A renewed study at Pena Blanca, Mexico, is under active consideration by the US DOE, for example.

Waste management organisations currently tend not to employ analogues in a systematic fashion to build repository system understanding and confidence in models and databases, although performance assessment aspects have been specifically evaluated in the Palmottu natural analogue project in Finland (EC 1996). Rather, analogues are used to provide information on specific issues and phenomena. The value of natural analogue arguments in the public debate is acknowledged; this has led different programmes to produce videos and brochures on the subject.

Appendix 4

OVERVIEW OF RESPONSES: PERFORMANCE AND SAFETY ASSESSMENT

A4.1   Progress in performance and safety assessment

A summary of performance assessments carried out since 1990, including a brief statement of purpose, is provided in Table A4.1.

Table A4.1.   **Recent (last 10 years) performance and safety assessments**
(partly based on Table 2 in NEA 1997)

Organisation	Purpose of performance/ Safety assessment	Documentation
ECN, RIVM, RGD, Netherlands	To study the option of waste disposal in a salt formation.	(Prij *et al.,* 1993).
AECL, Canada	To support decisions by regulatory agencies (or other bodies) on the future development of the nuclear waste programme.	The disposal of Canada's nuclear fuel waste: Post-closure assessment of a reference system (Goodwin *et al.,* 1994). The disposal of Canada's nuclear fuel waste: A study of the postclosure assessment of in-room emplacement of used CANDU fuel in copper containers in permeable plutonic rock (Goodwin e*t al.,* 1996).
NRI, RAWRA, Czech Republic	To evaluate the role of the barriers in a reference disposal system.	BAZ 97-02, The role of reference system in deep geological repository development (Konopaskova *et al,.* 1997)
Enresa, Spain	• To develop a performance assessment methodology. • To evaluate the role of the barriers in a granite host rock.	ENRESA-97 – Performance assessment of a spent fuel repository in granite (ENRESA, 1997).
	• To develop a performance assessment methodology. • To evaluate the role of the barriers in a clay host rock.	ENRESA-98 – Performance assessment of a spent fuel repository in clay (ENRESA, 1998).

Table A4.1 (cont'd).   **Recent (last 10 years) performance and safety assessments (partly based on Table 2 in NEA 1997)**

Organisation	Purpose of performance/ safety assessment	Documentation
European Commission	• To study the radiological impact of waste disposal in Boom Clay (medium-level waste).	PACOMA – Performance assessment of the geological disposal of medium-level and alpha waste in a clay formation in Belgium (Marivoet and Zeevaert, 1990).
	• To study the radiological impact of waste disposal in Boom Clay (spent fuel).	First Performance Assessment of the Disposal of Spent Fuel in a Clay Layer (Marivoet *et al.,* 1996).
		SPA (SPent Fuel Assessment) project (EC, 1998f)
	• Sensitivity analysis of geologic disposal systems in clay, granite and salt.	Evaluation of elements responsible for the effective engaged dose rates associated with the final storage of radioactive waste: EVEREST project (Cadelli, N. *et al.,* 1996, Marivoet, J. *et al.,* 1997).
Andra, France	• To do a preliminary assessment of the normal evolution scenario for each of the three potential deep sites	
GRS, Germany	• To study different disposal options.	Analysis of the long-term safety of disposal concepts with heat producing wastes (Buhmann *et al.,* 1991).
Nagra, Switzerland	• To support the selection of sites or geological media for a repository for high-level waste and long-lived intermediate-level waste. • To focus/summarise R&D status and provide support for continued R&D programme. • To provide a step in the ongoing process of evaluating the methodology for safety assessments.	Kristallin-I Safety Assessment Report (Nagra, 1994).
	• To assess intermediate-level waste from reprocessing.	Internal reports.

Table A4.1 (cont'd). **Recent (last 10 years) performance and safety assessments (partly based on Table 2 in NEA 1997)**

Organisation	Purpose of performance/ safety assessment	Documentation
Nagra, Switzerland	• To support site selection for a repository for low- and intermediate-level waste.	(Nagra, 1993)
	• To support a general license application for a repository for low- and intermediate-level waste.	(Nagra, 1994a)
HMIP, United Kingdom	• Trial of time-dependent PRA capability based on analysis of hypothetical L/ILW repositories at the Harwell site	Dry Run 3: Trial assessment of underground disposal based on probabilistic risk analysis (Sumerling ed., 1992)
Nirex, United Kingdom	• Part of the iterative development of a disposal concept.	Nirex 95: A preliminary analysis of the Groundwater Pathway for a Deep Repository at Sellafield (Nirex, 1995a).
	• To demonstration the generic capability developed to assess the performance of candidate sites. • To demonstrate the incorporation of site-characterisation and other R&D data into an assessment.	Nirex 97: An assessment of the post-closure performance of a Deep Waste repository at Sellafield (Nirex, 1997).
ONDRAF/SCK, Belgium	• To study the radiological impact of waste disposal in Boom Clay (medium- and high-level wastes).	UPDATING 1990 – Updating of the performance assessments of the geological disposal of high-level waste in the Boom Clay (Marivoet, 1991).
JNC, Japan	• To focus/summarise R&D status and provide support for continued R&D programme.	Research and development on geological disposal of high-level radioactive waste, First progress report (PNC, 1993).
	• To further demonstrate technical feasibility and reliability of geologic disposal concept. • To provide input for site selection and development of regulations.	The second progress report: H12 project for assessment of feasibility of HLW disposal in Japan (JNC, 1999).

Recent (last 10 years) performance and safety assessments (partly based on Table 2 in NEA 1997)

Organisation	Purpose of performance/ safety assessment	Documentation
SKB, Sweden	• To provide a step in the ongoing process of evaluating the methodology for safety assessments.	SKB-91, Final disposal of spent nuclear fuel. Importance of the bedrock for safety (SKB, 1992).
		SR-97 (to be published in Autumn 1999).
SKI, Sweden	• To develop regulatory review capacity (generic data).	SKI Project-90 (SKI, 1991).
	• To develop regulatory review capacity. • To demonstrate the incorporation of site-characterisation and other R&D data into an assessment.	SKI SITE 94 Deep Repository Performance Assessment Project (SKI, 1996).
POSIVA, Finland	• To support decisions by regulatory agencies (or other bodies) on the future development of the nuclear waste programme. • To support the selection of sites or geological media. • To focus/summarise R&D status and provide support for continued R&D pro-gramme.	TVO-92 safety analysis of spent fuel disposal (Vieno et al., 1992).
		TILA-96 safety assessment (Vieno and Nordman, 1996).
DOE/WIPP, USA	• To allow technical discussions with the regulatory body before the submission of a final license application. • To support the selection of sites or geological media.	Draft 40 CFR 191 Compliance Certification Application (DCCA) for the Waste Isolation Plant (SNL, 1995).
US NRC, USA	• To develop and demonstrate a perform-ance assessment methodology.	Initial demonstration of the NRC's capability to conduct a performance assessment for a high-level waste repository (Codell et al., 1992).

Table A4.1 (continued). **Recent (last 10 years) performance and safety assessments (partly based on Table 2 in NEA 1997)**

Organisation	Purpose of performance/ safety assessment	Documentation
US NRC, USA	• To develop regulatory review capacity.	NRC Iterative Performance Assessment Phase 2: Development of capabilities for review of a performance assessment for a high-level waste repository (Wescott *et al. eds.*, 1995).
	• To further develop regulatory review capacity. • To develop a regulation applicable to Yucca Mountain.	Iterative Performance Assessment Phase 3: status of activities (Manteufel & Baca, 1995).
US DOE/YMP, USA	• To provide feedback concerning the relative importance of site-specific characterisation and design information. • To develop more defensible assessment models for use in demonstration of compliance.	Total System Performance Assessment for Yucca Mountain-SNL Second Iteration (TSPA-1993) (Wilson *et al,*. 1994). Total System Performance Assessment – 1993: An Evaluation of the Potential Yucca Mountain Repository (Andrews *et al.*, 1994).
	• To support the selection of sites or geological media. • To provide decision makers with a "snapshot in time" on potential repository performance.	Total System Performance Assessment – 1995: An Evaluation of the Potential Yucca Mountain Repository (CRWMS M&O, 1995). Viability Assessment of a Repository at Yucca Mountain, Total System Performance Assessment (U.S. Department of Energy, 1998).

Note: PNC has undergone a reorganisation of its structure and, on 1st October 1998, was renamed the Japan Nuclear Cycle Development Institute (JNC).

The NEA's Working Group on Integrated Performance Assessments of Deep Repositories (NEA, 1997) reviewed a selection of performance assessments that were completed during the period 1991-1996. The group concluded that the NEA/IAEA/CEC Collective Opinion of 1991 (NEA, 1991) remains valid, i.e. they concluded that there are available today safety assessment methods which can be used to judge with sufficient reliability the performance of a deep repository. In particular, although the increased use of data from actual sites (and the increased detail in the specification of repository design) has presented new challenges and requires more resources than expended in earlier performance assessments, no new insurmountable problems in the use of performance assessment methods have been encountered (NEA, 1997).

Aspects of performance assessment where particular progress has been made include:

- the understanding of the performance of system components and their respective roles;

- treatment of uncertainty;

- the presentation of assessment findings;

- feedback to site selection, characterisation and repository design.

These aspects are discussed in more detail in the following sections. In addition, in response to the need to deal with large site datasets, progress has been made in more formal methods of data reduction for use in assessment models. The more sophisticated use of probabilistic codes has also been noted and there has been progress in understanding the strengths and weaknesses of probabilistic assessment techniques with respect to a deterministic approach (NEA, 1997). Some programmes now implement a combination of both methods.

Remaining areas where more work on performance assessment is seen as needed, or at least desirable, include:

- Sorption on canister corrosion products, which may contribute significantly to the safety of some repository concepts, but is not, in general, considered to be supported by sufficient data to include quantitatively in performance assessments.

- The treatment of climatic and geological events and changes (NEA, 1999c), although advances have been made in the quantitative assessment of the impact of climatic change (e.g. in the SKI SITE 94 performance assessment) and, in the USA, initial attempts have been made to quantify the effects of climate change, volcanic and seismic events on system performance.

- The treatment of coupled phenomena[6] (thermal, chemical, mechanical and hydrological), that may affect, for example, the early phase of heating and resaturation of a buffer, and also influence its long-term performance [FEBEX (EC, 1998e), PRACLAY (EUR 18047)].

- The impact of gas generated by repository materials, including waste, on system performance (e.g. the EC, PEGASUS project, EC, 1997b);

- The potential for the enhanced transport of radionuclides by colloids (e.g. U.S. Department of Energy,1998).

Whereas progress in these various aspects will obviously help to build confidence in the safety of a repository, the question arises as to when that confidence is sufficient. A degree of uncertainty is inevitable in quantitatively evaluating repository performance, particularly over very long time periods of hundreds of thousands or millions of years. Therefore, it has become increasingly recognised that a safety case in which there is overall confidence must be built on both of the following principles (NEA, 1999b).

- The repository system must exhibit the intrinsic robustness with respect to safety which is achieved through simple but robust design features, and suitable siting; i.e. the long-

---

6. Large-scale experiments are underway to develop understanding and test models of these phenomena (e.g. large-scale heater tests at Yucca Mountain and the FEBEX experiment at the Grimsel Test Site, and large-scale laboratory experiments in the JNC ENTRY programme, Japan).

term safety levels be relatively insensitive to uncertainties in the characteristics of any specific components.

- The safety assessment must exhibit the required degree of quality and reliability; this is achieved, for example, through the use of well tested performance assessment models and databases that incorporate conservative assumptions where there is uncertainty.

The former point can become ambiguous when advanced, more complex designs are proposed to provide additional assurance and safety. Do the extra safety margins which should result from the enhanced barrier system compensate for the additional uncertainties in evaluating the behaviour of these barriers and their interactions with other system components? Examples of such deliberations are the suggested use of drip shields or ceramic coatings to reduce the likelihood and amount of dripping advective water onto waste packages in the Yucca Mountain programme or the emplacement in a saturated environment of a complex, multicomponent engineered barrier system, including composite containers, specially treated bentonite buffers and intermediate layers of other materials. These concepts must be weighed against designs with simpler physical and chemical behaviour.

The engineered system can be made so robust that its performance minimises requirements on the site. Even in such a case, however, the site has always to provide an adequate geological environment – e.g. to provide protection from events at the surface and to guarantee low groundwater flow rates and favourable geochemical conditions. Analysing system performance and showing this analysis to be robust requires several types of calculations: system-level and subsystem level, probabilistic and deterministic, simplified and bounding calculations – as were all used in recent Canadian work on their major Environmental Impact Statement. Also, the application of more rigorous quality assurance (QA) procedures for R&D, assessment decisions, control of input/output datasets and code development has led to increased confidence in the findings of performance assessment. Many programmes use independent peer review within their QA procedures. NEA peer reviews have been conducted for the Canadian Nuclear Fuel Waste Management Concept, for the US Waste Isolation Pilot Plant (WIPP), and for the SKI SITE 94 Deep Repository Performance Assessment Project. In recent work, computer operating environments have even been explicitly designed to support the QA process for performance assessment calculations, ensuring reproducibility and traceability of results (e.g. the CAPASA system of JNC, Japan, Neyama et al,. 1998).

A4.2   Understanding of system components

Before progressing to analyses of the integrated repository system, it is imperative that one has an adequate, quantitative understanding of the behaviour of the system components. Performance or safety assessments consider the following broad components of a typical repository design:

(i)   the waste form itself;

(ii) the engineered structures around the waste, which may include, for example, high-integrity canisters and backfill materials;

(iii) the host rock in which the wastes are emplaced and surrounding geological units;

(iv) the biosphere.

Elements (i) and (ii) are commonly referred to as the "near field"[7] and (iii) as the "far field" or geosphere.

---

7.   The term "near field" is sometimes taken to include those parts of the host rock that are perturbed by the

These elements each have associated with them one or more safety functions. The function of the geological barrier, for example can be twofold. First a good geological environment will indeed function as a barrier to nuclide release and transport if it performs in the way expected. Much less is required of the geology, on the other hand, if its only function is to complement the engineered barrier functions by providing a stable, protected environment for the engineered system over long times.

Individual barriers can also have more than one safety function. For example, a steel canister can provide complete containment of radionuclides for an initial period, while, at later times, canister corrosion products give rise to chemical conditions that favour retention within the backfill. Thus, the various elements of a repository are to be viewed as complementary, rather than independent. This current view contrasts with the traditional description of repositories as systems of multiple, independent barriers, using, for example, a "Russian doll" analogy. It is thus important that the elements are compatible, e.g. geochemical compatibility between backfill materials and the geological environment. Many national and international programmes have sought, through performance assessment, to enhance their understanding of these safety functions and their relative importance.

At the level of individual safety barrier performance, significant advances can be identified in:

- the more realistic modelling of the processes that may lead to canister degradation and failure (e.g. the UK Nirex RARECAN model (Nirex, 1995b), ONDRAF/SCK, JNC, Nagra, SKI, US DOE and Enresa experimental and modelling studies of corrosion mechanisms and canister lifetime, Canadian AECL work);

- the understanding of waste-form degradation processes (e.g. experimental work on glass dissolution by Belgium, Japan and Switzerland; work by Enresa, Canada, Germany and the US on spent-fuel leaching processes under laboratory conditions);

- the modelling of early canister failures (SKB, 1997; US DOE, 1998);

- the more sophisticated use of geochemical codes and data to simulate pore-water composition and evolution (e.g. Nagra, the UK Nirex HARPHRQ model (Nirex, 1996) or the US DOE's EQ3 and 6 suite of models) and to arrive at element speciation and solubility equilibria (NEA, 1997);

- the effects on the properties of the near field and far field of the thermal history of the repository (ESF and URL and others, although the desirability of further work in this area has been noted);

- the use of three-dimensional models of groundwater flow, including density and transient effects, and the use of spatially-variable models of hydrogeological media, based on site data (NEA, 1997; e.g. mixed deterministic/stochastic discrete fracture network models in the SKI SITE 94 performance assessment, three-dimensional models of groundwater flow over a range of scales by Nagra and dual permeability modelling in the USA);

- the modelling of geosphere transport in fractured and unsaturated media and the testing of models against available field data (including the incorporation of colloid-facilitated radionuclide transport, for example by the US DOE and by Nagra);

presence of the engineered structures: e.g. the excavation-disturbed zone. In other cases, it is taken to include those parts of the rock that directly affect the engineered barrier system.

- better founded models of particular processes (e.g. volcanism and its effects, treatment of colloids, gas-mediated releases) (NEA, 1997) and of interactions between repository components (e.g. ONDRAF work on the interaction between vitrified waste, clay backfill and the host rock and work by the US NRC and AECL on the interaction between waste packages and the near-field environment).

Regarding the geosphere, one key issue for performance assessment is the extent to which site characterisation can provide data that can give confidence (or support analyses to demonstrate) that the required functions of this component of the repository system will be realised. The related use of performance assessment to guide field and laboratory programmes is addressed below.

Overall, confidence in both near-field and geosphere transport models, and the level of realism that such modelling can achieve, has increased over the last decade. Large-scale experiments have been performed that test models of the near field and hydrogeological aspects of the geosphere. Progress in the use of observations of natural systems to evaluate the isolation potential of the host rock has been noted, for example, in Germany, Switzerland and Canada, and there are ongoing EC projects on paleohydrogeology, including EQUIP and PAGEPA (EC, 1998f). The uncertainties associated with geosphere models are, however, generally still viewed as relatively high, due to the typical heterogeneity and variability of repository host rocks, to the difficulty of predicting the evolution of the host-rock environment over very long times, and to the fact that the models are less amenable to testing at relevant spatial and temporal scales. Furthermore, there is a limit to the extent to which a host rock can be characterised without perturbing or destroying its favourable characteristics. As a result, many recent performance assessments have emphasised the role of the near field in providing safety, with the geosphere being treated in a relatively simple and conservative fashion, at least in the earlier phases of development.

Nevertheless, there have been significant advances in treatment of the geosphere barrier. Understanding of the nature of water conducting features over a range of scales has definitely increased. Furthermore, some progress in the more realistic incorporation of natural variability in models of geosphere transport (e.g. more explicitly modelled "fast-path flow" in fractures by the US DOE, by JNC and by Nagra) and of the near-field/geosphere interface has been accomplished (e.g. in the recent work of Posiva). In a fractured rock, although there may be little prospect of excluding the possibility that a few "fast channels" for radionuclide transport may intersect the repository, a large part of the repository may be poorly connected, or not connected, to these channels. More reliable modelling of the near-field/geosphere interface may follow from the enhanced understanding of the excavation-disturbed zone that is expected to result from work at underground rock laboratories.

In the case of the biosphere, where there are uncertainties that are, in practice, impossible to quantify, there is a trend towards the use of a small number of stylised treatments, as discussed in the following section.

A4.3   Treatment of uncertainty

The treatment of uncertainties in parameter values etc. is a well established technique in modelling. In analysing the long-term behaviour of repositories, the more intractable problem has always been the handling of uncertainty in conceptual models of how the system might evolve in the future. Progress has been made in recent years on the more systematic and transparent treatment of features, events and processes (FEPs) which can determine future evolution and on handling uncertainty in the description of initial system behaviour. This includes the comprehensive exploration of the potential evolution of the environment at hand, followed by, or in concert with, identification of relevant FEPs indicated in part by those potential future environments. As part of the FEP cataloguing and evaluation process, there is a need to check that all site-specifically relevant FEPs are considered, e.g. through an audit of project-specific FEP lists against the International FEP database (NEA, 1999), and to track decisions on treatment and/or incorporation of FEPs into assessment models (NEA, 1997). The process thus includes the examination of a broad range of evolution scenarios, the evaluation of conceptual uncertainties by applying different models to the same phenomenon and sensitivity analysis of parameter uncertainty.

Some uncertainties may be identified that are, in practice, impossible to quantify and to reduce. Examples are uncertainties regarding:

- inadvertent human intrusion (although the likelihood of this event can perhaps be reduced by the choice of host rock type and by the selection of the site within that host rock);

- the evolution of the surface environment or biosphere;

- the "standard-man" assumption for the dose-effect relationship.

Nevertheless, as noted in (NEA, 1997), these issues must be addressed in performance assessments.

Rather than attempting to model in detail, or estimate in detail the likelihood of, these aspects of the system, performance assessors often choose to acknowledge that uncertainties make this impractical and then treat the corresponding part of the repository system in a stylised or simplified manner. The performance assessor makes a set of assumptions regarding these aspects, based on, for example, regulatory guidance, expert elicitation and, where this is available, international consensus. Examples are:

- the definition of a set of stylised human-intrusion scenarios;

- stylised biospheres (e.g. BIOMOVS II Reference Biosphere Methodology, van Dorp *et al.* 1999).

The acceptability of stylised treatments cannot be decided by the implementer alone, although the implementer may contribute with suggestions on how to treat such situations. If results for comparison with regulatory criteria are being calculated, then the regulator and other relevant decision makers will judge whether a stylisation is acceptable or not, and, in some cases, may provide guidance on what is, or is not, an acceptable approach. Confidence in the assessment capability need not be compromised provided that the documentation clearly acknowledges that these assumptions have been made and that, due to the presence of irreducible uncertainties, the results of the assessments are to be viewed as indicators of system behaviour based on these assumptions, rather than as predictions of consequences that will actually occur in the future. This approach has been taken by the USNRC in its recent draft regulations for a safety standard at Yucca Mountain (USNRC).

A4.4   Presentation of assessment findings

Although the exact contents of a performance assessment report will be dependent on a number of programme-specific and practical constraints, a consensus has emerged as to the broad elements that such a report should contain. There may, indeed, be advantages to a degree of standardisation in the presentation of results, both between assessments by a particular organisation and also between organisations. A regulatory compliance analysis, for example, is typically written in such a way as to ease the regulator's burden of performing an in-depth critical review.

A recommended list of eighteen elements (or topics) is given in (NEA, 1997), based on a review of recent performance assessments. This review also made suggestions as to the attributes that a performance assessment report should seek to achieve, such as:

- Traceability: an unambiguous and complete record of the decisions and assumptions made, and the models and data used, in arriving at a given set of results.

- Transparency: the clear reporting of a performance assessment, so that the audience can gain a good understanding of what has been done, what the results are, and why the results are as they are.

The US NRC is currently developing acceptance criteria for these, as well as specific technical content attributes that the staff will use to examine US DOE's license application. In Sweden, SKI has developed the concept of an Assessment Model Flowchart (the AMF (SKI, 1996)) as a tool for the traceable documentation of information processing and information transfer in a performance assessment.

The presentation of results has posed a number of problems to the organisations involved in performance and safety assessment. These are associated, for example, with the time-scales over which assessments are performed. Presentation is a particular problem where probabilistic assessment methods are employed: the large amounts of results generated by such methods must be distilled in such a way that the key findings are clearly shown.

Aspects where progress has been made include:

- The graphical representation of assessment results; a trend has emerged towards presentational methods that illustrate the performance of system components, as well as overall performance, and methods that illustrate where radionuclides reside within the system, as a function of time (e.g. the work of AECL, Nagra and DOE).

- The use of "insight" models, which capture and explain the behaviour of key radionuclides, based on simple physical and chemical principles (contributing to confidence in the correctness of more complex models and providing valuable presentation and communication tools).

- The production of specific reports tailored to different audiences, including separate summary reports for "programme managers" and for local communities, EIA-groups and the interested public.

- The incorporation of qualitative understanding in the argumentation of safety.

- The placing of assessment results in perspective with the risks associated with other human activities.

- The identification of the need for a statement of confidence (with supporting arguments) in the findings of a performance assessment (NEA, 1999b).

- The recognition that appropriate caveats should be placed on the results and conclusions of a performance assessment, acknowledging the limitations in the technical scope of the assessment and the potential impact of these limitations on the analysis.

Despite these specific advances, however, several organisations recognise the presentation of results, especially in view of the diverse audience, as an area where further work is required.

A4.5   Feedback to site selection, characterisation and repository design

Few performance assessments set themselves the specific aim of assisting in the process of site selection (exceptions are GSF-91 (Buhman *et al.*, 1991) and Nagra assessments for low- and intermediate-level waste). On the other hand, the use of performance assessment for optimising the programme of site characterisation and laboratory investigations and for evaluating and improving repository layout is being pursued by several organisations. The need for effective communication between those involved in performance assessment and those involved in site characterisation has been widely recognised and the close integration of geologists, hydrogeologists, designers and performance assessment modellers is now specifically aimed at by most disposal programmes. Such interaction serves to focus characterisation on safety-relevant issues and avoids the selection of sites and designs (and the development of models) for which the data necessary to demonstrate safety is likely to be unattainable. It was pointed out in (NEA, 1997) that, in many organisations, performance assessment personnel also participate in engineering design-evaluation work, although, in general, design selection is not seen to be as important as site evaluations in performance assessment studies. As projects approach licensing phases, however, there is an increasing need for a rigorous demonstration of the assumed function of the engineered system.

Appendix 5

REFERENCES FOR APPENDICES 2 TO 4

Andrews, R.W., Dale, T.F., and McNeish, J.A. (1994), *Total System Performance Assessment – 1993: An Evaluation of the Potential Yucca Mountain Repository*. B00000000-01717-2200-00099 Rev. 01. Las Vegas, Nevada: CRWMS (Civilian Radioactive Waste Management System) M&O (Management and Operating Contractor).

ARAP (1992), *Alligator Rivers Analogue Project, Final Report Volume 1, Summary of Findings*. An OECD/NEA International Project Managed by Australian Nuclear Science and Technology Organisation. ISBN 0-642-59927-0, also DOE/HMIP/RR/92/071 and SKI TR 92:20-1.

ASARR (1996), *Analogue Studies in the Alligator Rivers Region, Six Monthly Reports, 1 January to 30 June 1996 and 1 July to 31 December 1996*. Available from the project manager or from the NEA Secretariat.

Buhmann *et al.* (1991), *Analysis of the long-term safety of disposal concepts with heat producing radiactive wastes*. GSF, Braunschweig, Bericht 27/9, 1991 (in German).

Cadelli, N. *et al.* (1996), *Evaluation of elements responsible for the effective engaged dose rates associated with the final storage of radioactive waste: EVEREST project: Summary report*. EUR 17122.

CAN (1998), *Nuclear Fuel Waste Management and Disposal Concept, Report of the Nuclear Fuel Waste Management and Disposal Concept Environmental Assessment Panel, February 1998*. Minister of Public Works and Government Services Canada, EN-106-30/1-1998E.

CNE (1998), *Reflexions sur la réversibilité des stockages, Commission nationale d'évaluation, Juin 1998*. (Executive summary in English available: Thoughts on Retrievability).

Codell, R. B. et al. (1992), *Initial demonstration of the NRC's capability to conduct a performance assessment for a high-level waste repository*. U.S. Nuclear Regulatory Commission, NUREG-1327.

Council of Europe (1998), Parliamentary Assembly of the Council of Europe, Resolution 1157 (1998) Radioactive waste management.

Cramer, J.J. and Smellie, J.A.T. (1994), *Final report of the AECL/SKB Cigar Lake analog study*. Atomic Energy of Canada Limited Report, AECL-10851, COG-93-147, SKB TR 94-04.

CRWMS (Civilian Radioactive Waste Management System) M&O (Management and Operating Contractor) (1995), *Total System Performance Assessment – 1995: An Evaluation of the*

Potential Yucca Mountain Repository. B00000000-01717-2200-00136 REV 01. Las Vegas, Nevada: CRWMS M&O.

CUM (1997), Planning Inspector's report, Cumbria County Council Appeal by United Kingdom Nirex Limited, C. S. McDonald, File No. APP/H0900/A/94/247019, March 1997.

van Dorp, F., Egan, M., Kessler, J. H., Nilsson, S., Pinedo, P., Smith, G., Torres, C. (1999), "Biosphere modelling for the assessment of radioactive waste repositories; the development of a common basis by the BIOMOVS II reference biospheres working group". In *Journal of Environmental Radioactivity*, 42, 225-236.

EC (1995), *Radionuclide transport through the geosphere and biosphere, Review study of the project MIRAGE*. EUR 16489.

EC (1995a), *Testing and modelling of thermal, mechanical and hydrogeological properties of host rocks for deep geological disposal of radioactive waste, proceedings of a workshop, Brussels, 12- 13 January 1995*. EUR 16219.

EC (1996), *Sixth EC natural analogue working group meeting, Proceedings of an international workshop held in Santa Fe, New Mexico, USA, 12-16 September 1994*. EUR 16761.

EC (1997a), *Seventh EC natural analogue working group meeting, Proceedings of an international workshop held in Stein am Rhein, Switzerland, 28-30 October 1996*. EUR 17851.

EC (1997b), *Projects on the effects of gas in underground storage facilities for radioactive waste, (Pegasus project)*. EUR 18167 and EUR 16746, 16001, 15734, and 14816.

EC (1997c), *El Berrocal project, Characterization and validation of natural radionuclide migration processes under real conditions on the fissured granitic environment: Final report*. EUR 17478.

EC (1998), *OKLO working group, Proceedings of the first joint EC-CEA workshop on the OKLO-natural analogue Phase II project, held in Sitges, Spain, 18-20 June 1997*. EUR 18314.

EC (1998a), *Nuclear fission safety – Progress report 1997. Vol. 2: Radioactive waste management and disposal and decommissioning*. EUR 18322/2.

EC (1998b), *In situ testing in underground research laboratories for radioactive waste disposal. Proceedings of a cluster seminar held in Alden Biesen (B), 10-11 December 1997*. EUR 18323.

EC (1998c), *The Praclay project: Demonstration test on the Belgian disposal facility concept for high activity vitrified waste: Final report*. EUR 18047.

EC (1998d), "RESEAL: A large scale in situ demonstration for REpository SEALing in an argillaceous host rock". In *In situ testing in underground research laboratories for radioactive waste disposal, Proceedings of a cluster seminar held in Alden Biesen (B), 10-11 December 1997*. EUR 18323.

EC (1998e), *FEBEX Project: a Full scale Engineered Barriers EXperiment in crystaline rock, in In situ testing in underground research laboratories for radioactive waste disposal, Proceedings of a cluster seminar held in Alden Biesen (B), 10-11 December 1997.* EUR 18323.

EC (1998f), *Nuclear fission safety - Progress report 1997, Volume 2:Radioactive waste management and disposal and decommissioning.* EUR 18322/2. (Progress report 1998 under preparation for publication in the EUR series).

ENRESA (1997), *ENRESA-97 – Performance assessment of a spent fuel repository in granite.* Enresa Technical Publication 06/97 (in Spanish).

ENRESA (1998), *ENRESA-98 – Performance assessment of a spent fuel repository in clay.* Enresa Technical Publication (in preparation).

GER (1998), Gesetz über die friedliche Verwendung der Kernenergie and den Schutz gegen ihre Gefahren (Atomgesetz). Vom 23. Dezember 1959 (BGB1. IS. 814) in der Fassung der Bekanntmachung vom 15. Juli 1985 (BGB1. IS. 1565) (BGB1. III 751-1) zuletzt geändert durch Gesetz zur Änderung des Atomgesetzes und des Gesetzes über die Errichtung eines Bundesamtes für Strahlenschutz vom 6. April 1998 (BGB1. IS. 694).

Glassley, W. (1994), "Validation of hydrogeochemical codes using the New Zealand geothermal system". In *Proceedings of the Fifth CEC Natural Analogue Working Group (NAWG) meeting and Alligator Rivers Analogue Project (ARAP) Final Workshop, 5 –9 October 1992, Toledo, Spain.* EC-NAWG H.von Maravic and J. Smellie (Editors), 1994. EUR 15176 EN, Luxembourg .

Goodwin *et al.* (1994), *The disposal of Canada's nuclear fuel waste: Postclosure assessment of a reference system.* AECL-10717, COG-93-7.

Goodwin *et al.* (1996), *The disposal of Canada's nuclear fuel waste: A study of the postclosure assessment of in-room emplacement of used CANDU fuel in copper containers in permeable plutonic rock, Volume 5: Radiological assessment.* AECL-11494-5, COG-95-552-5.

Griffault, L. *et al.* (1996), *Acquisition et régulation de la chimie des eaux en milieu argileux pour le projet de stockage de déchets radioactifs en formation géologique : Projet "Archimède argile" : rapport final.* EUR 17454.

ICRP (1986), *ICRP Publication 46: Radiation Protection Principles for the Disposal of Solid Radioactive Waste.* Annals of the ICRP Vol 15/4.

ICRP (1993), *ICRP Publication 64: Protection from Potential Exposure: A Conceptual Framework. A Report of a Task Group of Committee 4 of the International Commission on Radiological Protection.* Annals of the ICRP Vol. 23/1.

ICRP (1998), *ICRP Publication 77: Radiological Protection Policy for the Disposal of Radioactive Waste.* Annals of the ICRP Vol. 27 Supplement.

ICRP, *Radiation Protection Recommendations as Applied to the Disposal of Long-Lived, Solid Radioactive Waste.* To be published by the International Commission on Radiological Protection, (in preparation).

IAEA (1997), Joint Convention on the Safety of Spent Fuel Management and on the Safety of Radioactive Waste Management (adopted on 5 September 1997 and opened for signature at the Headquarters of the IAEA).

IAEA (1998), *Technical, institutional and economic factors important for developing a multinational radioactive waste repository*. TECDOC-1021 (1998).

JNC (1999), *The Second Progress Report: H12 Project for Assessment of Feasibility of HLW Disposal in Japan*. (In preparation).

Kickmaier, W. & McKinley, I. (1997), "A review of research carried out in European rock laboratories". In *Nuclear Engineering and Design 176* , pp 75-81.

Konopaskova S., Pergl L. (1997), *Model Description of a Reference Disposal System*. Nuclear Research Institute (NRI), REZ, Czech Republic, BAZ 97-02, 90 pages (in Czech).

Kowalski, E., and Fritschi, M.(1996), "Swiss underground L/ILW repository Wellenberg after the negative vote of the canton." *IAEA Symposium on experience in the Planning and Operation of Low Level Waste Disposal Facilities, Vienna, 17-21 June 1996*. IAEA-SM-341/53.

Manteufel, R. D. & Baca, R. G. (1995), *Iterative Performance Assessment Phase 3: status of activities*. San Antonio, Texas, Center for Nuclear Waste Regulatory Analyses. CNWRA 95-007 (prepared for the U.S. Nuclear Regulatory Commission).

Marivoet, J. and Zeevaert, Th. (1990), *PACOMA – Performance assessment of the geological disposal of medium-level and alpha waste in a clay formation in Belgium*. EUR 13042 EN.

Marivoet, J (1991), *UPDATING 1990 – Updating of the performance assessments of the geological disposal of high-level waste in the Boom Clay*. SCK – ONDRAF/NIRAS report BLG 634.

Marivoet, J., Volckaert, G., Snyers, A. & Wibin, J (1996), *First Performance Assessment of the Disposal of Spent Fuel in a Clay Layer*. EUR 16752 EN.

Marivoet, J. et al.(1997), *Evaluation of elements responsible for the effective engaged dose rates associated with the final storage of radioactive waste: EVEREST project*. EUR 17449.

Miller, W., Alexander, R., Chapman, N., McKinley, I., Smellie, J.(1994), "Natural analogue studies in the geological disposal of radioactive wastes". In *Studies in Environmental Science 57*, Elsevier (Amsterdam, London, New York, Tokyo),1994 - also as: Nagra Technical Report 93-03, Nagra, Wettingen, Switzerland, 1994.

Murphy, W.M., Pearcy, E.C. and Pickett, D.A., (1997). "Natural analogue studies at Pena Blanca and Santorini". In: *Proceedings of the Seventh EC Natural Analogue Working Group (NAWG) Meeting, 28 – 30 October 1996, Stein am Rhein, Switzerland*. EC-NAWG H. von Maravic and J. Smellie (Editors), EUR 17851 EN, Luxembourg.

Nagra (1993), *Beurteilung der Langzeitsicherheit des Endlagers SMA an Standort Wellenberg (Gemeinde Wolfenschiessen, NW)*. Nagra Technical Report Series NTB 93-26, Nagra, Wettingen, Switzerland.

Nagra (1994), *Kristallin-I safety assessment report*. Nagra Technical Report Series NTB 93-22, Nagra, Wettingen, Switzerland.

Nagra (1994a), *Bericht zur Langzeitsicherheit des Endlagers SMA am Standort Wellenberg*. Nagra Technical Report Series NTB 94-06, Nagra, Wettingen, Switzerland.

NEA (1991), *Disposal of radioactive waste: can long-term safety be evaluated? A Collective Opinion of the Radioactive Waste Management Committee of the OECD Nuclear Energy Agency and the International Radioactive Waste Management Advisory Committee of the International Atomic Energy Agency, endorsed by the Experts for the Community Plan of Action in the Field of Radioactive Waste Management of the Commission of the European Communities*. OECD Nuclear Energy Agency, Paris.

NEA (1993), *The Cost of High-Level Waste Disposal in Geological Repositories: An Analysis of Factors Affecting Cost Estimates*. OECD Nuclear Energy, Paris.

NEA (1997), *Lessons learnt from ten performance assessment studies*. OECD Nuclear Energy Agency, Paris.

NEA (1997a), *Field Tracer Experiments: Role in the Prediction of Radionuclide Migration, Synthesis and proceedings of a workshop held in Cologne, Germany, 28-30 August 1996*. OECD Nuclear Energy Agency, Paris.

NEA (1998), *Modelling the Effects of Spatial Variability on Radionuclide Migration, Synthesis and proceedings of a workshop held in Paris, France, 9-11 June 1997*. OECD Nuclear Energy Agency, Paris.

NEA (1999), *An International Database of Features, Events and Processes*. OECD Nuclear Energy Agency, Paris, in press.

NEA (1999a), *Low-Level Radioactive Waste Repositories: An Analysis of Costs*. OECD Nuclear Energy, Paris.

NEA (1999b), *Confidence in the Evaluation of Safety of Deep Geological Repositories*. OECD Nuclear Energy, Paris, (in preparation).

NEA (1999c), *Characterisation of Water-Conducting Features and their Representation in Models of Radionuclide Migration, Synthesis and proceedings of a workshop held in Barcelona, Spain, 10-12 June 1998*. OECD Nuclear Energy Agency, Paris, in press.

Neyama *et al.* (1998), "Quality assurance program with computer-oriented management system for performance assessment." In *High-level Radioactive Waste Management, Proceedings of the Eighth International Conference, Las Vegas, Nevada, May 11-14, 1998*. American Nuclear Society, Inc., La Grange Park, Illinois 60526 USA.

Nirex (1995a), *Nirex 95 - A preliminary analysis of the Groundwater Pathway for a Deep Repository at Sellafield*. Nirex Science Report S/95/012.

Nirex (1995b), *The Development and Application of RARECAN*. F. M. Porter and A. V. Chambers, Nirex Safety Series Report NSS/R396.

Nirex (1996), *HARPHRQ: A Computer program for Geochemical Modelling*. A. Haworth, T. G. Heath and C. J. Tweed, Nirex Report NSS/R380.

Nirex (1997), *Nirex 97 - An assessment of the post-closure performance of a Deep Waste repository at Sellafield*. Nirex Science Report S/97/012.

PNC (1993), *First Progress Report on Research and Development on Geological Disposal of High-Level Radioactive Waste*. PNC TN1410 93-059.

Prij, J. *et al.* (1993), *PROSA – Probabilistic Safety Assessment, Final Report*. ECN, RIVM, RGD, Petten.

SKB (1992), *Final Disposal of Spent Nuclear Fuel; Importance of the Bedrock for Safety*. SKB Technical Report 92-20, The Swedish Nuclear Fuel and Waste Management Co.

SKB (1997), *Assessment of a spent fuel disposal canister. Assessment studies for a copper canister with a cast steel inner component*. A.E. Bond, A.R. Hoch, G.D. Jones, A.J. Tomaczyk, R.M. Wiggin, W.J. Worraker; AEA Technology, Harwell, UK; SKB TR 97-19.

SKB (1998), *Parameters of importance to determine during geoscientific site investigations*. J. Andersson, K-E. Alméen, L.O.Ericsson, A. Fredriksson, F. Karlsson, R. Stanfors, A. Ström; SKB TR 98-02.

SKB (1998a), *Geoscientific evaluation factors and criteria for siting and site evaluation*. A Ström, J. Andersson, K-E. Elméen, C Svemar, LO Ericsson; SKB-R-9907.

SKB (1998b), *Maqarin Natural Analogue Study: Phase III, vol. I and II*. Smellie, J.A.T., Conterra AB (ed.); SKB TR 98-04.

SKI (1991), *SKI Project-90*. SKI Technical report 91:23, Swedish Nuclear Power Inspectorate, Stockholm.

SKI (1996), *SKI SITE-94 Deep Repository Performance Assessment Project*. SKI Report 96:36, Swedish Nuclear Power Inspectorate.

Smellie, J.A.T., Karlsson, F. and Alexander, R., (1997), "Natural analogue studies: present status and performance assessment implications". In *Journal of Contaminant Hydrology 26* (1997) 3-17.

SNL (1995), *Draft 40 CFR 191 Compliance Certification Application (DCCA) for the Waste Isolation Plant*. SNL, 1995.

US DOE (1997), *Linking Lagacies; Connecting he Cold War Nuclear Weapons Production Processes to their Environmental Consequences*. DOE/EM-0319, 1997.

US DOE (1998), *Viability Assessment of a Repository at Yucca Mountain, Total System Performance Assessment*. DOE/RW-0508 Volume 3, North Las Vegas, Nevada: DOE Office of Civilian Radioactive Waste Management, Yucca Mountain Site Characterization Office.

USNRC (1998), *Draft proposed U.S. Nuclear Regulatory Commission, Regulations for the disposal of high-level radioactive wastes in the proposed geologic repositry at Yucca Mountain, Nevada, USA*. U. S. NRC, Draft proposed to CFR PART 63, 1998.

Vieno, T., Hautojärvi, A., Koskinen, L., Nordman, H. (1992), *TVO-92 Safety Analysis of Spent Fuel Disposal*. Nuclear Waste Commission of Finnish Power Companies, Report YJT-92-33E.

Vieno, T., Nordman, H. (1996), *Interim report on safety assessment of spent fuel disposal TILA-96*. Helsinki: Posiva, 1996. 176 s. (POSIVA 96-17). ISBN 951-652-016-2.

Wescott, R. G. *et al. eds*. (1995), *NRC Iterative Performance Assessment Phase 2: Development of capabilities for review of a performance assessment for a high-level waste repository*. U.S. Nuclear Regulatory Commission, NUREG-1464.

Wilson, M.L., Gauthier, J.H., Barnard, R.W., Barr, G.E., Dockery, H.A., Dunn, E., Eaton, R.R., Guerin, D.C., Lu, N., Martinez, M.J., Nilson, R., Rautman, C.A., Robey, T.H., Ross, B., Ryder, E.E., Schenker, A.R., Shannon, S.A., Skinner, L.H., Halsey, W.G., Gansemer, J.D., Lewis, L.C., Lamont, A.D., Triay, I.R., Meijer, A., and Morris, D.E. (1994), *Total System Performance Assessment for Yucca Mountain-SNL Second Iteration (TSPA-1993)*. SAND93-2675. Albuquerque, New Mexico: Sandia National Laboratories.

Wolfsberg, A. V. *et al.* (1999), "Use of Chlorine-36 and other Geochemical Data to Test a Groundwater Flow Model for Yucca Mountain, Nevada". In *Use of Hydrogeochemical Information in Testing Groundwater Flow Models, Technical summary and proceedings of a workshop held in Borgholm, Sweden, 1-3 September 1997*, OECD Nuclear Energy Agency, Paris.

Appendix 6

OVERVIEW OF RESPONSES: COMMUNICATION AND PERCEPTION

In the questionnaire, the importance of communication and its effect upon subjective judgements on the status of geologic disposal was recognised by eliciting information on:

- How disposal implementers or regulators communicate with the public?

- Whether the efforts made to communicate have led to any perceptible changes in attitudes?

- Which events or developments have had most impact on attitudes?

In a final section of the questionnaire, the subjective opinions of the respondents themselves were sought, asking them to subjectively address global influencing factors under the following titles

- the most positive developments over the last decade;

- the most negative developments in the same timeframe;

- the future actions which would most likely advance the status of geologic disposal programmes.

The results of these forays into the field of public communication and perception are summarised in the present appendix.

A6.1   Communication

Public Participation in Decision Making

An initial question is what constitutes public participation. Many levels of involvement of stakeholders in disposal programmes are possible and different countries have differing approaches. The most direct form of public involvement is for the public to directly participate in the decision-making process leading to the acceptance of a specific proposal by the proponent or the granting of a license by the authorities. Only Switzerland mentions the possibility given to opponents of a project to initiate public referenda on such issues, but referenda of a consultative nature have been carried out also in countries like Sweden.

Many countries, however, have formalised processes for allowing public input to decision making in the waste disposal area. Direct representation in public hearings is relatively uncommon, but consultation in written form is common. A requirement for formalised consultation with the public is sometimes associated with a generally applicable Environmental Protection Act or similar law, e.g.

in Hungary, USA, Canada and in countries that have implemented EC directives requiring that an Environmental Impact Assessment be carried out for facilities whose construction or operation might result in a significant impact upon the environment. The EIA is carried out by the operator or proponent of the facility and made available for public information and comment.

Several countries have no formalised requirement for public involvement specifically in waste disposal projects. In almost all countries, however, public opinion is nevertheless actively elicited by governments, regulators and proponents in a less formalised way to allow:

- the proponent to adapt his project to take account of the views of the public;

- the licensing authority to factor in comments by the general public before proposing regulatory guidance;

- the government authorities to consider public opinion before making final licensing type decisions;

The responses give a clear picture that there is today no shortage of means by which the public can give input to the planning and implementation processes for waste repositories. Even where there are no strong constraints on proponents or regulators to consult closely with the public, the common tendency in such a controversial field is to seek a dialogue. The following section illustrates the lengths which are gone to by all parties to encourage an informed debate.

Communicating with Stakeholders

Most organisations consider that it is their duty to reach out to the:

- general public;
- political decision makers;
- local affected public and their representatives;
- wider scientific and technical audiences.

The organisations also indicate that they fulfil this duty by a wide variety of means (Table A6.1).

Of the respondents, only the EC mentions the availability of an Internet site, although many organisations do have a home page which includes also the opportunity for public response. The role of the Internet as a fast, real-time means to reach out to, and be reached by, several audiences will obviously grow and this should be noted by waste management organisations.

Between regulators and implementers there is clearly a difference in their own perceived roles as information sources. Some authorities and regulators (e.g. in Canada) are extremely wary of being seen as nuclear proponents and prefer to interact more with decision makers. Other regulators like SKI in Sweden or EPA in the USA, whilst still avoiding a pro-nuclear stance, take a very proactive role in factually informing the public and encouraging a dialogue emphasising in particular the role of independent regulation in assuring public safety.

Table A6.1. **Means adopted by waste management organisations to reach out to different audiences**

Audience	Implemented means
General public	Periodic publications, leaflets, videos, exhibitions, visitor centres, site visits, Internet web-sites with an e-mail question and answer capability, participation in public debates, open house tours, ...
Local public	All of above, mixed committees of representatives of local public and waste management organisations, talks at schools, personal contacts, ...
Political decision makers	Written advice, contributions to advisory panels
Wider scientific community	Annual R&D reports, publication in scientific journals and conferences, lectures at universities, peer review groups, ...

The importance attached by almost all respondents to communication is underlined by the fact that most have staff dedicated directly to this function. Mostly there are relatively few purely public relations staff. A common view is that their core efforts should be augmented by technical specialists working in other parts of the organisation. In a few cases, all of the communications work is entrusted to technical staff. In general, there is agreement that both technical and non-technical staff should be employed in order to provide effective communication of a type and depth which will differ from audience to audience.

Key information to be communicated

An interesting set of responses were elicited by the question on key information to be communicated. The variety of messages that need to be given to different audiences is exemplified in Table A6.2. The most repeated key messages, however, were as follows:

- Geologic disposal when properly done is safe; only very low levels of risk are involved – but this is difficult to communicate convincingly to a wider audience.

- There is a pragmatic need for disposal. Radioactive waste exists and must be disposed of independently of the future of nuclear energy. The obvious coupling to the future of nuclear energy, however, ensures continued opposition from specific groups.

- The solution to waste disposal should be developed by the present generation. Interim storage or transmutation are not seen by the waste community as an alternative to disposal; the first offers a postponement, the second offers a modification of the inventory to be disposed.

Interestingly, while most organisations do mention the need to pass on information on the low level of risk involved, they do not mention efforts to make the public aware of other types of risk that society is taking or is willing to take. This may reflect evidence that the public does not place much weight on comparisons of risks with widely different origins but requires more direct assurance in specific cases.

Table A6.2.   **Key messages**

- Technical safety of disposal; Low level of risk

- Need for disposal

- Environmental and ethical reasons for disposal

- Geologic disposal is an internationally accepted route

- Reasonableness and fairness of a stepwise development process

- Importance of an independent regulator

- Quality, and ethical stature of implementing regulations

- Local benefits to host communities

- Disposal is in principale reversible

The key messages to pass on are very basic and are common to all. One obvious question arising concerns the effectiveness of information campaigns in the past. Why is every organisation still trying to communicate the same information? Have the messages been provided in the past in an ineffective way?

A6.2   Perceptions

Changes in attitudes during the last decade

The questionnaire tried to gather information on the changes in public opinion over the last ten years and on the effectiveness of communication efforts in altering this opinion relative to the influences of external events. Several responses elicited on overall changes in attitudes during the last ten years indicated that, although general acceptance of nuclear power may have edged higher in their countries, similar progress can not be claimed for the disposal of long-lived waste. There has been an increased awareness of the issue of final disposal and there have been movements towards more public involvement in the siting process. This has tended to lengthen the process, not least because groups opposing waste disposal have become more organised and more capable of reaching audiences such as the local public and the media. The arguments that they present are also increasing in sophistication. This increases the need for implementers and, where appropriate, regulators to communicate early and openly if they are to get their own messages across.

At first sight, there is not a perception that the technical advances discussed in earlier chapters have moved countries much closer to the implementation of geologic disposal than they were ten years ago. One significant exception is in Finland, where the implementation of the EIA process and of two LLW repositories seems to have had positive impacts. A more differentiated analysis indicates that some other programmes also moved ahead in the 10 year period considered – mainly by implementing facilities (e.g. SFR in Sweden), by moving to site specific geologic investigations (e.g. Belgium, UK, USA, Switzerland), or by passing an important licensing step (e.g. WIPP in the US). But a distinct perception is that overall recent setbacks have reduced or eliminated apparent gains made earlier.

Attitudes are certainly strongly affected by nuclear events occurring in the world in general, and, even more strongly, in the national disposal programmes. Chernobyl is the single most cited cause for looking with heightened suspicion not only at nuclear power but also at nuclear waste management and disposal. At a national level, incidents not directly related to disposal can also negatively affect attitudes. Examples are the Monju and Tokai accidents in Japan and the public revelation of contaminated transport casks in European countries. Specific setbacks in disposal programmes also raise the level of public concern. Examples here are the rejection of the planning application for the RCF in the UK, the negative Swiss referendum on Wellenberg, the negative outcome of referendums in two "pre-study communities" in Sweden, the dropping of the Spanish siting programme and the Canadian expert panel recommendation not to move into a siting phase at the present time. All of these negative events,[1] on the one hand, mirror the sceptical public attitude to disposal and, on the other hand, themselves feed back to increase public concern.

In most of the disposal programme setbacks, the problems have centred around the contentious issue of siting facilities. Specific failed attempts at siting have often failed based more on lack of sufficient local trust in the implementers than on real technical deficiencies in the project work. A rounded package including demonstrably sound technical work, good communications, appropriate compensation schemes and personal trust seems to be necessary to win acceptance, especially in the face of dedicated opposition by opponent groups. The failures in some countries, together with the fact that disposal is easily postponed by prolonging storage times, have discouraged other programmes from even proceeding to the siting phase.

Contribution of waste management organisations to attitude changes

One hope of waste management organisations is that the public may ultimately gain confidence through the demonstrated willingness of implementers to take all their appropriate concerns seriously. In many countries, such as Sweden, the UK, Canada and Switzerland, events have demonstrated that the public does have a voice that is indeed heeded in the debate on repository implementation.

Given fact, it is crucial that developers succeed in communicating their messages to the public. A corresponding question on the effectiveness of public information efforts produced, however, from most repository implementers or regulators the sobering response that their information

---

1. The term "positive" is used in this report to describe situations where disposal programmes are moving forward. The terms "negative" or "setback" describe situations where siting decisions are being re-examined and programmes are not progressing towards disposal as planned. These terms are used informally in implementing organisations, though not necessarily by regulators, and represent a subjective judgement. The overall governmental or societal judgement may, in fact, be that the current re-evaluations of siting schedules are "positive" developments. If new schedules and siting decisions are made, and governments become active proponents of geologic disposal, it may be that current "negative" developments will be "positive" at some point in the future, when decisions are taken and supported.

and communication efforts are judged to have had very modest positive influence on public awareness or acceptance of disposal projects.

The apparent impasse reached by the limited ability of the waste management community to encourage public consensus on moving ahead with disposal justifies fully the concerns of the RWMC and others. It also argues for increased attention to be devoted by the community to the issues involved, even if these issues do not strictly fall within traditional areas of science and engineering.

Progress towards geologic disposal in the last ten years

The final block of questions was aimed specifically at soliciting the subjective opinions of the respondents from within the waste community on the most positive and most negative developments towards implementation of waste disposal. In considering those positive and negative developments, some respondents focused on national developments, while some cited both national and international events and developments (Tables A6.3 and A6.4).

Table A6.3. **Most frequently cited international developments, and examples of national developments, positively influencing progress in deep disposal**

Positive developments and events in national programmes
Actual implementation of national waste management facilities.
Beginning construction of facilities that may eventually become repositories, e.g. sinking of Gorleben shafts in Germany, and construction of access tunnels at Yucca Mountain.
Progress in site characterisation and safety assessment studies, such as issuance of the Canadian Environmental Impact Statement;
Publication of milestone national safety and feasibility studies.
Government policy statements endorsing progress towards geologic disposal or defining waste management strategy.
Regulatory reform and/or issue of regulatory guidance.
Local and regional campaigns allowing direct communication and increased public participation.
Recognition of the need for deep disposal for wastes other than the highly radioactive waste from commercial nuclear power.
Positive developments and events with international significance
Progress towards implementation of waste management facilities in other countries (e.g. WIPP certification was the most commonly quoted example of progress).
Progress in site characterisation and safety assessment studies in other countries.
Maturing of techniques of site characterisation and safety assessment which allow more robust analyses.
International co-operation in R&D and notably in underground rock laboratory projects.
Growing international collaboration and consensus.
The work of the international organisations, to foster collaboration and consensus (most commonly mentioned was the issue of the NEA/EC/IAEA Collective Opinion of 1991 and the NEA Collective Opinion of 1995).
Increased efforts on evaluating the influence on disposal programmes of alternative strategies, e.g. partitioning and transmutation (P&T), direct disposal of SF, use of MOX, retrievability.
Growing awareness of environmental hazards posing much greater problems and receiving very much less attention than radioactive waste disposal (especially Greenhouse effect).

Table A6.4.   **Most frequently cited international developments, and examples of national developments negatively influencing progress in deep disposal**.

Negative developments and events in national programmes
Setbacks in national siting and investigation programmes, e.g. the rejection of the RCF in the UK, referenda in Switzerland and Sweden and the conclusions recommending postponement and modification by the Canadian evaluation panel.
The successful opposition by pressure groups and the local population against proposed site characterisation and test drilling studies.
The underestimation of the force of public opinion and of the ability of opposition groups to manipulate this.
Lack of governmental will or legal framework at the national level.
The coupling of waste management issues in a broader debate about the appropriate future of nuclear energy.
Budgetary pressures that tend to limit research, favour lower cost options or delay disposal projects, cited in Belgium and the USA.
Negative developments and events with international significance
Setbacks in siting and investigation programmes of other countries.
The successful opposition by pressure groups and the local population against proposed site characterisation and test drilling studies.
Lack of perceived urgency about developing permanent repositories, coupled with belief that a better technical solution might be available if we wait (e.g. the revival of P&T as an excuse for delaying work on geologic disposal); discussions on retrievability/long-term monitoring concepts.
Continuing legacy of distrust of nuclear industry from past closed information policies and nuclear accidents, e.g. Chernobyl and Three Mile Island, news of massive remediation requirements in former Soviet states and, to a lesser extent, in the US.
The decline in interest in many western countries for the nuclear energy option in general.
International disagreements on the scientific basis of safety and contradictory messages from the scientists on the safety of disposal.

Negative developments in foreign countries were perceived to affect national and international programmes more than positive events. Perhaps this reflects the relatively restrained, and nationally limited, information on successes normally transmitted by implementing organisations and the necessarily restrained outreach efforts of regulators. By contrast, opponents are becoming more sophisticated and effective in their outreach efforts, globally.

Licensing of waste disposal facilities, e.g. the VLJ facility in Finland, are, of course, cited as national developments but are also cited as having positive influence by some respondents in other countries. In particular, several respondents noted the licensing of the WIPP facility as a significant world-wide positive development. Correspondingly, setbacks in investigation programmes, e.g. the rejection of the RCF proposals in the UK, and votes against developments and studies, e.g. in Sweden and Switzerland, have direct impact in national programmes but were also cited as negative factors by some respondents in other countries. It may be the case, however, that positive or negative developments in one country have a less direct effect on decision makers or public opinion in other countries than on the morale of implementer organisations. Invariably press reports of negative occurrences in the nuclear industry elsewhere are referenced to similar local facilities or plans to give them a local-interest perspective. International consensus amongst governments and the technical community on the desirability of waste disposal is seen as a crucial positive factor, e.g. as communicated in the NEA Collective Opinions of 1991 and 1995. Conversely, the growing pan-national trend for re-examination of options such as partitioning and transmutation, and prolonged storage, is generally seen as a negative factor, although in some countries, e.g. the Netherlands and Belgium, this was noted as a positive development within the national context. Some respondents identified a loss of interest in western countries for nuclear energy in general and/or a lack of urgency about developing permanent repositories as reasons for this increased emphasis on P&T and storage, and in some responses financial factors were implicated.

Continuing international collaboration in R&D related to geologic disposal, notably, co-operation in underground rock laboratory projects is seen as an important positive factor. Development and maturing of techniques for site characterisation and safety assessment is also cited as positive by several respondents. International disagreements on the scientific basis of safety and contradictory messages from the scientists on the safety of disposal were, however, seen as negative.

Lack of governmental will to implement unpopular developments or a deficient legal or regulatory framework at the national level is cited as a negative factor by a few respondents. Correspondingly, positive government statements and development of a proper and implementable regulatory framework and guidance are stated as positive factors in those countries where these apply.

A continuing legacy of distrust of the nuclear industry from policies that, in the past, limited the release of information, giving the public an impression of secrecy, and from past nuclear accidents, especially Chernobyl, were identified as negative factors in public opinion. In Finland, however, the lack of any recent accident was cited as a positive factor.

Requirements for progress in the next ten years

In considering the developments required to improve progress towards geologic disposal over the next 10 years, again national and international developments were cited. The developments hoped for by the waste management community can be grouped into broad categories (Table A 6.5).

Many respondents again emphasised the importance of international events and developments, especially the positive influence that the development of repositories in other countries might have as a working example of demonstrably safe facilities. This indicates, perhaps, that most respondents (implementers and regulators) believe they are working on the right lines within their own national contexts, but a groundswell of international progress would lift them more easily over the various national hurdles. Alternatively, respondents see a lack of progress internationally as feeding the opposition to national programmes with powerful public arguments against the feasibility of disposal projects. Where respondents did cite specific national requirements for progressing repository programmes, the emphasis was on policy and organisational aspects, and the mechanisms for gaining public acceptance for current technical solutions, rather than development of improved technical solutions.

Table A6.5. **Most frequently cited developments required to improve progress towards geologic disposal over the next 10 years**

Developments which would most help progress in waste disposal

Visible progress in implementation

Successful siting of geologic repositories in one or several countries - these can serve as examples of facilities which have been demonstrated in a public process to be safe.

Agreement of local municipalities/communities to let an implementer conduct surface-based site investigations or start underground excavations (for URLs or for characterising proposed disposal facilities).

Continued work towards repositories at Gorsleben and Yucca Mountain and installation of underground laboratories in France.

Placement of transuranic waste in the WIPP facility.

Increased public acceptance

Setting up an effective system for securing public acceptance that geologic disposal at a site selected in accordance with a defined process is an appropriate course of action.

Providing both the public and legislators with sufficient information defining the urgency associated with developing permanent repositories, coupled with belief that improved technical solutions can be integrated into the system as it progresses.

Development of simple, understandable internationally harmonised safety and technological criteria to guide the radioactive waste management programmes at national and international levels.

Confidence building in scientific basis and safety assessments of disposal (e.g. by demonstration-scale experiments in URLs).

Demonstration of feasibility of manufacturing and quality control of engineered barriers.

Promoting awareness and better understanding of legitimate comparative risks involved with all technologies, especially with respect to waste management.

Improved development framework

Designation of organisations to fulfil various roles and development of associated legal frameworks and decision making processes.

Development of simple, understandable, perhaps internationally harmonised, safety and technological criteria to guide the radioactive waste management programmes at national and international levels.

Reduction of political implications/influences on repository project realisation as well as on repository operation.

Policy or strategy statements

Clear government endorsement of deep disposal as being the sustainable solution for management of long-lived nuclear waste.

Re-affirmation at the technical level of achievable safety of repositories.

Further aids to progress

An initiative to explore the possibilities of a regional or multinational repository.

An objective study of alternative strategic and technical options.

Increased funding (mentioned only by few programmes).

Few respondents stressed the role of government, most presumably believing that government support was either established or would flow from public acceptance. Where government action was sought, there was some disagreement on the precise type of involvement favoured. For example, both a Canadian and a UK respondent favoured strong, direct government endorsement of deep disposal as being the sustainable solution for management of long-lived nuclear waste and a clear procedural framework allowing for stepwise endorsement by government at key stages in the programme, whereas a German respondent wished for sound political decisions combined with a reduction of political influence on repository project realisation.

The main developments which would be welcomed by respondents can be summarised as follows:

- At an international level: a clear consensus on the need for geologic disposal, planned and implemented in a stepwise manner; confidence in the technical feasibility and safety of the concept; and that consensus and confidence to be expressed in the development of one or more successfully operating repositories.

- At a national level: clear procedures for staged siting studies and repository development, and methods for communicating effectively and for gaining public acceptance in the stepwise development of appropriate national solutions.

Appendix 7

KEY TEXTS OF PUBLISHED COLLECTIVE OPINIONS OF THE NEA

From: *Technical Appraisal of the Current Situation in the Field of Radioactive Waste Management, 1985*

SUMMARY AND CONCLUSIONS

Industrial activities are regarded as safe even though a small risk always exists. The philosophy of radiation protection accepts this and recognises that some level of risk will also be associated with safe radioactive waste management. Therefore the objective of radioactive waste management is to look for a strategy which, taken as a whole, is considered safe and provides an acceptable balance of all the radiological, technical, social, political and economic considerations. The RWMC's appraisal underlines the need for such a balance while concentrating on radiological and technical factors, particularly on the long term safety aspects of radioactive waste management.

The fundamental conclusion is that detailed short and long term safety assessments can now be made which give confidence that radiation protection objectives can be met with currently available technology for most waste types, and at a cost which is only a small fraction of the overall cost of nuclear-generated power. The other main conclusions on both the short and long term aspects of radioactive waste management are as follows:

On the short term, which covers the operational life of waste management facilities and any period of institutional control:

- Radiological protection objectives can be consistently met during the operation of a facility and for as long afterwards as controls are maintained for all currently used or envisage radioactive waste management concepts.

- Storage can be relied upon for all waste types as an interim measure, as long as appropriate surveillance and monitoring is provided.

- While high priority is currently given to the full development and early demonstration of disposal concepts, there is no urgency to dispose of the small volumes of high level radioactive waste and spent fuel currently accumulated, as they can continue to be stored safely until disposal is judged appropriate.

On the long term, which covers the post-institutional control period:

- Specific long term radiation protection objectives for radioactive waste disposal have been developed to provide a basis for judging the radiological acceptability of disposal practices or developing specific criteria for individual waste types.

- Predictive risk assessment methodologies have been developed for the assessment of the long term safety of disposal systems.

- There is a high degree of confidence in the ability to design and operate disposal systems in deep geological structures which will assure long term isolation for high level waste or spent fuel and meet the relevant long term safety objectives.

- While the short term aspects of uranium mine and mill tailings can be safely managed, there remains some concern in the long term about human intrusion into tailings or their possible misuse, and long term requirements need to be established.

An overall impression of optimism and confidence prevails from the RWMC's appraisal. It results from the substantial body of scientific and technical evidence from past and ongoing studies and R+D activities as well as from the experience already available. At the same time, it is recognised that :

- R+D will have to continue, notably to fill remaining gaps for particular options, to collect site-specific data and to refine safety studies;

- periodic reassessments of waste management practices and policies will have to be made to take account of evolving knowledge; and

- quality control at all stages is an essential nuclear safety requirements and it will have to be applied throughout the whole sequence of waste management activities.

In this situation, the RWMC considers that a step by step approach to the application of waste management technologies as they become viable on an industrial scale, is both justifiable and safe.

From: *Disposal of Radioactive Waste: Can Long-term Safety Be Evaluated? An International Collective Opinion, Paris 1991*

CONCLUSIONS: THE INTERNATIONAL VIEW

International co-operation–through information exchange and joint projects–plays a substantial role in the development of methods for safety assessment. In particular, international co-operation promotes periodic and systematic reviews of the state-of-the-art in this field, and contributes to informed and objective debate among specialists.

Following such a review, the NEA Radioactive Waste Management Committee and the IAEA International Radioactive Waste Management Advisory Committee

- *recognise* that a correct and sufficient understanding of proposed disposal systems is a basic prerequisite for conducting meaningful safety assessments;

- *note* that the collection and evaluation of data from proposed disposal sites are the major tasks on which further progress is needed;

- *acknowledge* that significant progress in the ability to conduct safety assessment has been made;

- *acknowledge* the quantitative safety assessments will always be complemented by qualitative evidence and

- *note* that safety assessment methods can and will be further developed as a result of ongoing research work.

Keeping these considerations in mind, the two Committees

- confirm that safety assessment methods are available today to evaluate adequately the potential long-term radiological impacts of a carefully designed radioactive waste disposal system on humans and the environment; and

- consider that appropriate use of safety assessment methods, coupled with sufficient information from proposed disposal sites, can provide the technical basis to decide whether specific disposal systems would offer to society a satisfactory level of safety for both current and future generations.

This collective Opinion is endorsed by the CEC Experts for the Community Plan of Action in the Field of Radioactive Waste Management.

From: *The Management of Long-lived Radioactive Waste. The Environmental and Ethical Basis of Geological Disposal. A Collective Opinion of the NEA Radioactive Waste Management Committee, Paris, 1995*

As part of its continuing review of the general situation in the field of radioactive waste management, and with particular reference to the extensive discussions at the recent NEA Workshop on Environmental and Ethical Aspects of Radioactive Waste Disposal [1], the RWMC reassessed the basis for the geological disposal strategy from an environmental and ethical perspective at its Special Session in March 1995. In particular, the RWMC focussed its attention on fairness and equity considerations:

- between generations (intergenerational equity), concerning the responsibilities of current generations who might be leaving potential risks and burdens to future generations; and

- within contemporary generations (intragenerational equity), concerning the balance of resource allocation and the involvement of various sections of contemporary society in a fair and open decision-making process related to the waste management solutions to be implemented.

After a careful review of the environmental and ethical issues, as presented later and discussed in detail in the proceedings of the NEA Workshop, the members of the NEA Radioactive Waste Management Committee:

- *consider* that the ethical principles of intergenerational and intragenerational equity must be taken into account in assessing the acceptability of strategies for the long-term management of radioactive wastes;

- *consider* that from an ethical standpoint, including long-term safety considerations, our responsibilities to future generations are better discharged by a strategy of final disposal than by reliance on stores which require surveillance, bequeath long-term responsibilities of care, and may in due course be neglected by future societies whose structural stability should not be presumed;

- *note* that, after consideration of the options for achieving the required degree of isolation of such wastes from the biosphere, geological disposal is currently the most favoured strategy;

- *believe* that the strategy of geological disposal of long-lived radioactive wastes:
  - takes intergenerational equity issues into account, notably by applying the same standards of risk in the far future as it does to the present, and by limiting the liabilities bequeathed to future generations; and

  - takes intragenerational equity issues into account, notably by proposing implementation through an incremental process over several decades, considering the results of scientific progress; this process will allow consultation with interested parties, including the public, at all stages;

- *note* that the geological disposal concept does not require deliberate provision for retrieval of wastes from the repository, but that even after closure it would not be impossible to retrieve the wastes, albeit at a cost;

- *caution* that, in pursuing the reduction of risk from a geological disposal strategy for radioactive wastes, current generations should keep in perspective the resource deployment in other areas where there is potential for greater reduction of risks to humans or the environment, and consider whether resources may be used more effectively elsewhere.

Keeping these considerations in mind, the Committee members:

- *confirm* that the geological disposal strategy can be designed and implemented in a manner that is sensitive and responsive to fundamental ethical and environmental considerations;
- *conclude* that it is justified, both environmentally and ethically, to continue development of geological repositories for those long-lived radioactive wastes which should be isolated from the biosphere for more than a few hundred years; and
- *conclude* that stepwise implementation of plans for geological disposal leaves open the possibility of adaptation, in the light of scientific progress and social acceptability, over several decades, and does not exclude the possibility that other options could be developed at a later stage.

Appendix 8

LIST OF ACRONYMS

AECL   Atomic Energy of Canada Limited, Canada

ANDRA National Agency for Radioactive Waste Management, France

EC; CEC European Commission; Commission of the European Community

ECN   Netherlands Energy Research Center, Petten, the Netherlands

ENRESA National Waste Management Company, Spain

GEOTRAP NEA International Project on the Transport of Radionuclides in Geologic, Heterogeneous Media

GRS   Company for Nuclear and Industrial Plant Safety, Germany

HLW   High-Level Waste

HMIP   Her Majesty's Inspectorate of Pollution (now part of the Environment Agency for England and Wales), United Kingdom

IAEA   International Atomic Energy Agency, Vienna, Austria

ICRP   International Commission on Radiological Protection

IPAG   NEA Working Group on the Integrated Performance Assessments of Deep Repositories

JNC   Japan Nuclear Cycle Development Institute (former PNC)

NAGRA National Cooperative for the Disposal of Radioactive Waste, Switzerland

OECD/NEA Nuclear Energy Agency of the Organisation for Economic Co-operation and Development, Paris, France

Nirex   UK Nirex Ltd, United Kingdom

NRI   Nuclear Research Institute, Rez, Czech Republic

ONDRAF/NIRAS National Organization for Radioactive Wastes and Fissile Materials, Belgium

SCK/CEN	Nuclear Energy Research Center, Mol, Belgium
PA	Performance Assessment
PAAG	NEA Performance Assessment Advisory Group
RAWRA	Radioactive Waste Repository Authority, Czech Republic
RGD	Dutch Geological Survey (now part of TNO), the Netherlands
RIVM	National Institute of Public Health and Environment Protection, the Netherlands
RWMC	NEA Radioactive Waste Management Committee
SEDE	NEA Co-ordinating Group on Site Evaluation and Design of Experiments for Radioactive Waste Disposal
SKB	Nuclear Fuel and Waste Management Company, Sweden
SKI	Nuclear Power Inspectorate, Sweden
URL	Underground Research Laboratory
US DOE	Department of Energy, USA
US NRC	Nuclear Regulatory Commission, USA
WIPP	Waste Isolation Pilot Plant, Carlsbad, USA
YMP	Yucca Mountain Project, USA

ALSO AVAILABLE

NEA Publications of General Interest

1998 Annual Report (1999)   *Free: paper or Web.*

NEA Newsletter
ISSN 1016-5398   Yearly subscription: FF 240 US$ 45 DM 75 £ 26 ¥ 4 800

Radiation in Perspective – Applications, Risks and Protection (1997)
ISBN 92-64-15483-3   Price: FF 135 US$ 27 DM 40 £ 17 ¥ 2 850

Radioactive Waste Management in Perspective (1996)
ISBN 92-64-14692-X   Price: FF 310 US$ 63 DM 89 £ 44

Radioactive Waste Management Programmes in OECD/NEA Member countries (1998)
ISBN 92-64-16033-7   Price: FF 195 US$ 33 DM 58 £ 20 ¥ 4 150

Radioactive Waste Management

Fluid Flow through Faults and Fractures in Argillaceous Formations (1998)
ISBN 92-64-16021-3   Price: FF 400 US$ 67 DM 119 £ 41 ¥ 8 100

Water-conducting Features in Radionuclide Migration
ISBN 92-64-17124-X   Price: FF 600 US$ 96 DM 180 £ 60 ¥ 11 600

Lessons Learnt from Ten Performance Assessment Studies (1997)   *Free: paper or Web.*

Nuclear Waste Bulletin (1998)   *Free: paper or Web.*

Progress Towards Geologic Disposal of Radioactive Waste: Where Do We Stand? (1999)
                                                                        Free: paper or Web.

*Confidence in the Long-term Safety of Deep Geological Repositories –Its Development
and Communication*   *Free: paper or Web.*

Order form on reverse side.

105

ORDER FORM

OECD Nuclear Energy Agency, 12 boulevard des Iles, F-92130 Issy-les-Moulineaux, France
Tel. 33 (0)1 45 24 10 10, Fax 33 (0)1 45 24 11 10, E-mail: nea@nea.fr, Internet: http://www.nea.fr

Qty	Title	ISBN	Price	Amount
		Postage fees*		
		Total		

*European Union: FF 15 – Other countries: FF 20

❏ Payment enclosed (cheque or money order payable to OECD Publications).

Charge my credit card   ❏ VISA   ❏ Mastercard   ❏ Eurocard   ❏ American Express

(N.B.: You will be charged in French francs).

Card No.	Expiration date	Signature
Name		
Address	Country	
Telephone	Fax	
E-mail		

OECD PUBLICATIONS, 2, rue André-Pascal, 75775 PARIS CEDEX 16
PRINTED IN FRANCE
(66 1999 17 1 P) ISBN 92-64-17194-0 – No. 51101 1999